Analytical Molecular Biology
Quality and Validation

Analytical Molecular Biology
Quality and Validation

Edited by

Ginny C. Saunders and Helen C. Parkes
Laboratory of the Government Chemist, Teddington, UK

With a Foreword by S. B. Primrose

*Setting standards
in analytical science*

VALID ANALYTICAL MEASUREMENT

ROYAL SOCIETY OF CHEMISTRY

ISBN 0-85404-472-8

A catalogue record for this book is available from the British Library

Published for Laboratory of the Government Chemist by The Royal Society of Chemistry,
Thomas Graham House, Science Park, Milton Road, Cambridge CB4 0WF, UK

For further information see our web site at www.rsc.org

Typeset by Paston PrePress Ltd, Beccles, Suffolk
Printed by Redwood Books Ltd, Trowbridge, Wiltshire, UK

Foreword

The organisation and sequence of the DNA in any organism defines the biological and biochemical properties of that organism and its absolute taxonomic position. It also serves as the most accurate identifier of that organism. Closely related organisms have only small differences in DNA sequence whereas distantly related organisms have major differences in sequence and organisation.

Twenty-five years ago the ability to analyse sequences at the gross level, never mind the determination of the absolute order of bases in a short sequence, was restricted to a small and select group of scientists. These scientists truly had skills in DNA biochemistry and they were highly respected and much sought after. Today, analysis of DNA sequences is a commonplace technology practised by scientists from a diversity of disciplines, many of whom have had no training in practical biochemistry. As a consequence, DNA analysis finds application in a wide range of pure and applied sciences. Perhaps the biggest growth area is in molecular diagnostics: pathogen detection, forensics (establishment of guilt, identification of perpetrators of crimes), clinical genetics (detection of inherited disorders), oncology (pre-disposition to cancer), food analysis (detection of adulteration), archaeology (identity of fossil remains) and geography (population movements).

The pervasive spread of DNA technology is, to a large extent, due to the rapid growth in companies offering a wide range of convenience kits for gene manipulation and analysis. Use of these kits almost guarantees that the laboratory worker will get a result. Indeed, users often are not aware that the procedures they are following require exquisite attention to detail. The quality of each and every component must be assessed and, because everything is conducted on a micro-scale, great care must be taken to ensure that the correct quantities are used. This is what kit suppliers do for the customer. There are two disadvantages associated with this widespread use of these kits. First, many people use them without understanding the biochemistry involved. Second, because most of the kits come with robust protocols, many users are experimentally sloppy. One consequence is that the user fails to get any result. Alternatively, even if they do get a result, is it the correct result? The issue is of particular relevance in diagnostic applications, especially ones where nucleic acid amplification has been employed. It cannot be over-emphasised.

This manual has been prepared mainly by staff at the Laboratory of the Government Chemist. Since the Laboratory is the premier UK centre for materials analysis, the authors are only too well aware of the need for precision and accuracy in molecular DNA analysis. They also have a deep understanding of the factors leading to unwanted variance. Their objective in writing this manual is to provide an authoritative guide to all the key procedures to ensure that *practitioners not only get a result, but that they get the right result.*

S. B. Primrose

Contents

Chapter 6   Inhibitors and Enhancers of PCR   81

Jane Bickley and Daniel Hopkins

Acknowledgements

We wish to thank all of our colleagues who have contributed to this manual. Our particular thanks go both to the authors of individual chapters, who are therefore acknowledged directly, and Alan Calder, Nicholas Wisbey, Helen Dewhurst and Nigel Burns of the LGC for their contributions of additional experimental data in support of the issues raised. Dr David Holcombe (LGC), Professor Jack Firth (VAM steering group), Dr Jane Shallcross (CAMR), Dr James Walker (UDL) and Dr Graham Reed (NMSPU) have also been kind enough to assist by commenting on working drafts during the preparation of this manuscript. Johanne Cornett has additionally played a critical role in the final stages of editing and preparation of this manual. The work described was supported under contract with the Department of Trade and Industry as part of the National Measurement System Valid Analytical Measurement Programme. The ongoing support of Professor Sandy Primrose for the application of the VAM Initiative to molecular biology is also gratefully acknowledged.

Ginny C. Saunders
Helen C. Parkes

Abbreviations

A	adenine
A_{260}	absorbance at 260 nm
A_{280}	absorbance at 280 nm
ABI	Applied Biosystems International
AFLP	amplified fragment length polymorphism
AP PCR	arbitrarily primed PCR
bp	base pair
BSA	bovine serum albumin
c7dGTP	7-deaza-2'-deoxyguanosine triphosphate
C	cytosine
CAPS	cleaved amplified polymorphic sequence
cDNA	complementary DNA
CPD-Star™	disodium 4-chloro-3-(methoxyspiro{1,2-dioxetane-3,2'-(5'-chloro)tricyclo[3.3.1.1^{3,7}]decan}-4-yl)phenyl phosphate
CRM	certified reference material
CSPD®	disodium 3-(4-methoxyspiro{1,2-dioxetane-3,2'-(5'-chloro)tricyclo[3.3.1.1^{3,7}]decan}-4-yl)phenyl phosphate
CTAB	cetyltrimethylammonium bromide
DAF	DNA amplification fingerprinting
dATP	deoxyadenosine triphosphate
dCTP	deoxycytidine triphosphate
ddH$_2$O	double distilled H$_2$O
DGGE	denaturing gradient gel electrophoresis
dGTP	deoxyguanosine triphosphate
DI	detectability index
DIECA	diethyldithiocarbamic acid
DIG	digoxigenin
DMF	dimethylformamide
DMSO	dimethyl sulfoxide
DNA	deoxyribonucleic acid
DNase	deoxyribonuclease
dNTP	deoxyribonucleoside triphosphate
dsDNA	double-stranded DNA
DTI	Department of Trade and Industry
DTT	dithiothreitol
dTTP	deoxythymidine triphosphate

dUTP	deoxyuridine triphosphate
EDTA	ethylenediaminetetraacetic acid
EGTA	ethylene glycol bis(β-aminoethyl ether)tetraacetic acid
EQA	external quality assessment
EtBr	ethidium bromide
G	guanine
GLP	Good Laboratory Practice
GM	genetically modified
HMW	high molecular weight
IAA	isoamyl alcohol (isopentyl alcohol)
IgG	immunoglobulin G
ISO	International Organisation for Standardisation
kbp	kilobase pair
mPCR	multiplex PCR
mRNA	messenger RNA
MWM	molecular weight marker
NAMAS	National Accreditation of Measurement and Sampling
NIBSC	National Institute for Biological Standards and Control
NIST	National Institute of Standards and Technology
NP40	nonidet P40
OD	optical density
PAGE	polyacrylamide gel electrophoresis
PBS	phosphate buffered saline
PCR	polymerase chain reaction
PE	Perkin Elmer
PEG	poly(ethylene glycol)
PFGE	pulsed field gel electrophoresis
PT	proficiency testing
PVP	poly(vinylpyrrolidone)
QA	quality assurance
QC	quality control
QPCR	quantitative PCR
RAPD	randomly amplified polymorphic DNA
RFLP	restriction fragment length polymorphism
RM	reference material
RNA	ribonucleic acid
RNase	ribonuclease
rRNA	ribosomal RNA
RT PCR	reverse transcriptase PCR
SDS	sodium dodecyl sulfate
SGM	second generation multiplex
SSC	salt sodium citrate
SSCP	single-strand conformation polymorphism
ssDNA	single-stranded DNA
STR	short tandem repeat
T	thymine

TE	tris-EDTA buffer
T_m	melting temperature
TMAC	tetramethylammonium chloride
Tris	tris(hydroxymethyl)aminomethane
TSGE	temperature sweep gel electrophoresis
U	uracil
UDG	uracil DNA glycosylase
UKAS	UK Accreditation Service
UV	ultraviolet
VAM	valid analytical measurement
WHO	World Health Organisation

Weights, volumes and concentrations

M	molar
mol	mole
g	gram
l	litre

Prefixes

m	milli (10^{-3})
μ	micro (10^{-6})
n	nano (10^{-9})
p	pico (10^{-12})
f	femto (10^{-15})
a	atto (10^{-18})

CHAPTER 1

An Introduction to Analytical Molecular Biology

GINNY C. SAUNDERS AND HELEN C. PARKES

1.1 Introduction

DNA technology is having a revolutionary effect on a host of industrial and regulatory sectors. The pace of fundamental innovation in the biosciences shows no signs of abating and continues to reveal new commercial opportunities in both biotechnology and analytical molecular biology. Healthcare, pharmaceutical production, diagnostics, agriculture, animal husbandry, food and forensic analysis are just a few areas where DNA technology is significantly changing the way industry and regulators operate. Clearly, this rapidly developing technology offers tremendous advantages and benefits to bioanalysis with respect to increased scope of application, detection limits, speed, cost and specificity. However, in order to capture and utilise these advantages, there is an urgent need for parallel validation of the analytical techniques employed in DNA-based measurements. The cost of employing invalid or flawed DNA technology would be enormous and highly damaging, both in terms of public perception and financial investment.

Analytical molecular biology has been typically developed in the academic and medical research environments. Here, priorities are understandably concerned with innovation, rather than consideration being given to the more routine applicability, reliability and reproducibility of the methods. Evaluation of these factors and further method validation is therefore an absolute prerequisite for the successful move of techniques from the research laboratory to the analytical laboratory.

Limited discussion at scientific fora has been paid to questioning the validity of DNA-based measurements, despite growing commercial and public activity in these areas. There are possibly three main reasons for the lack of research and debate into the validity of these measurements. First, the excitement of being able to measure where no-one has measured before can lead to an enthusiastic rush of application. Second, regulation of the analysis is generally carried out in-house and not through performance standards set by the larger analytical community. Finally, there is a lack of reference samples such as key analytes contained in complex matrices necessary for the critical comparison of analytical approaches.

This manual aims to introduce and address quality and validation issues that arise in the application of DNA technology and, hopefully, offers a basis for further discussion and debate within the bioanalytical community.

1.2   What is Analysis, Why is it Undertaken?

Analysis is usually initiated, proposed or commissioned by a customer, who can be a private individual or company, public organisation or law enforcement agency such as the police force or trading standard office. Analysis of a material or matrix is undertaken to examine one or more of its constituent parts or analytes. Analytical data are required as an independent source of information in order for the customer to gauge a situation, interpret evidence, decide whether action is required or to ascertain whether certain regulations are being adhered to. The data obtained from analysis are therefore required in a variety of forms:

- Qualitative — confirmation of the presence of an analyte
- Semi-quantitative — provides an estimate of analyte concentration
- Quantitative — provides a well defined value for the amount of analyte

There are also various types of analyses that can be undertaken, each offering different discriminatory powers. These are summarised in Table 1.1.

Table 1.1 *Different types of analyses that can be undertaken and the information that can be obtained*

Types of analyses	Data obtained
Detection (screening)	Positive/negative result, fast, high throughput, lower cost, qualitative
Confirmatory (could be monitoring or characterisation type of analysis)	Quantitative or qualitative, highly specific, used following a presumptive positive in screening analysis
Monitoring (surveillance)	Can be quantitative or qualitative, high specificity, used for the detection of change of an analyte
Characterisation (identification/profiling/diagnosis/genotyping)	Qualitative, various levels of specificity and validity, used to discriminate at various levels (can involve a quantitative description of the characterisation)
Reference	Fully validated, definitive measurement

Analysis should not be viewed as a straightforward exercise or in any sense mundane due to its sometimes routine nature. In reality, analytical methodologies are frequently made up of a complex and evolving mixture of techniques, where specific applications or samples demand appropriate adaptations. A seemingly straightforward implementation of the methodologies and generation of data could arise from either careful and considered planning and validation, showing a dedication to producing quality analytical data, or a complete lack of all the aforementioned qualities. In the second case, implementation appears simple as the task has not been undertaken with due consideration or care. Chapter 2 discusses how to obtain the former scenario and avoid the latter.

1.3   DNA, a Universal Biological Analyte

Increasingly high expectations of public health and general quality of life has led to a greater need for the detection and analysis of biological materials. Detection of human, animal, food and environmental pathogens can all inform public health policy. The advent of biological methodologies such as DNA forensics has revolutionised the analysis of scene of crime evidence and provided a valuable tool for law enforcement agencies such as the police, trading standard offices and wildlife protection organisations. Molecular genetic tests have allowed pre-natal detection of genetic diseases and can detect gene mutations which may inform a change of lifestyle.

In spite of the vast variety and complexity of biological materials (matrices and organisms), they share a host of common biomolecules, of which nucleic acids form a major group. Deoxyribonucleic Acid (DNA) is an ideal universal analyte for biological methodologies. It is the genetic material of the majority of forms of life and an identical copy of the genome is contained within nearly every cell of an organism. The DNA of an individual is unique (with the exception of homozygous twins) with respect to the sequential order of the four base constituents, making it an indisputable marker for identification purposes. A genome consists of both highly conserved regions of sequence such as genes and variable, non-conserved regions. Comparable DNA sequences show more similarity between closely related individuals or species and less similarity between distant relatives. Both non-conserved and highly conserved regions of a genome are exploited in analytical molecular biology to detect similarities or differences (known as DNA polymorphisms) of a DNA sequence.

The use of nucleic acids, particularly DNA, as an analyte offers unparalleled sensitivity to biological detection and characterisation techniques. Theoretically, using the polymerase chain reaction (PCR),[1,2] a single copy of a gene can be detected. In the field of bacterial detection and identification, DNA technology is, in many cases, offering faster analysis times than comparable classical methodologies such as plate culture detections. DNA is also more resistant to degradation than RNA or protein molecules, an important factor when selecting an analyte from highly processed or aged samples.

1.4   Sectoral Applications of Analytical Molecular Biology Techniques

Listed in Table 1.2 is a summary of current applications of analytical molecular biological methodologies. The range is so vast that these techniques could well touch everyone's life at some time or another and go some way to maintaining the current standard of living expected in the Western world.

Table 1.2 *Sectoral applications of analytical molecular biology techniques*

Sector	Example of sectoral application
Agriculture	Pathogen detection, plant breeding programmes, GM crop detection, cultivar identification
Animal husbandry	Identification of viral, fungal and bacteriological infections, progression of infection, assessment of treatment Design of breeding programmes through genetic characterisation Sex identification of animals and birds
Archaeology	Phylogenetics (the study of relationships and evolution), familial analysis, species identification
Clinical/healthcare	Genetic disease diagnosis, progression of disease, assessment of treatment, linkage analysis, pre-natal diagnosis Identification of viral, fungal and bacteriological infections, progression of infection, assessment of treatment and epidemiological studies Examination of archival clinical samples
Ecology	Measurement of biodiversity, sex identification, investigation of symbiotic interactions
Environment	Pathogen detection for environmental legislation
Food	Pathogen detection, product/species authentication, adulteration detection, GM food detection
Forensic science	Individual and familial identification
Law enforcement	Trading standards, e.g. detection of adulteration in food, drinks and fibres Immigration, i.e. familial analysis Wildlife protection, e.g. detection of wild birds/animals taken from the wild through familial analysis
Research	Phylogenetics (the study of relationships and evolution) Genome sequencing projects

1.5   Challenges of DNA Analysis

DNA analysis does, however, have its own challenges. Some major concerns arise from the analysis of 'real' samples, as in typical industrial and enforcement situations where non-ideal samples are the norm. Such samples originate from a variety of sectoral applications such as forensic, food or environment, where the

DNA analyte may be in association with an organic matrix, for example a blood stain on cotton fibre, *Listeria* spp. in milk or *Legionella* spp. in water.

Some of the challenging situations that exist in the application of DNA technologies are:

- *Low concentration of analyte compared to matrix.* This has lead to the development of sophisticated DNA extraction and amplification methodologies to selectively isolate and concentrate the analyte of interest. Examples include low level detection of environmental and food pathogens.
- *The varied and complex biological or chemical matrices* that are the source of the nucleic acid to be analysed can make DNA extraction a difficult undertaking. Complex chemical or biochemical components of a matrix, such as naturally occurring secondary compounds, can interfere with enzyme activity and can cause total inhibition of biological reactions such as PCR and restriction enzyme digests.
- *DNA degradation due to a sample being subjected to harsh conditions.* These include industrial processing such as freezing, dying, heating, grinding, tanning, drying and forms of weathering such as those caused by the sun or rain. Such conditions may be in addition to the ageing of a sample, all of which can cause physical degradation of the DNA analyte.
- *Biological contamination of the sample* can mean that nucleic acids from a variety of sources are present, perhaps due to environmental insult (e.g. bacterial or fungal contamination) or scene of crime samples containing bodily fluid from both the victim and the criminal. Endogenous or exogenous (i.e. from contaminating microorganisms) DNases can cause DNA degradation.
- *Degradation of matrix components* can sometimes produce breakdown products, such as polyphenols, that cause the degradation of nucleic acids.
- *Limited availability of a sample.* This may be because the sample represents a unique moment in time or is limited by quantity.
- *Lack of suitable controls.* There are very few characterised reference samples that can be employed to ensure the accurate calibration of equipment, the correct handling of samples or the applicability of methodologies.

It is partly due to the challenges listed here that there is a wide gap between molecular biological technique development and analytical application, leaving the transition from research to routine somewhat problematic. In order for a technique to become readily accepted as an analytical tool, confidence must be gained in the performance of the technique. An application must appear robust enough to avoid the production of erroneous results and be resistant to small changes in one or more of its parameters.

1.6   Key Techniques in Analytical Molecular Biology

From the wide range of molecular biology techniques available, only a selection is commonly employed in analysis (Table 1.3). Other techniques, such as cloning and transformation, are perhaps more widely employed in biotechnological applications and more state of the art techniques are most likely to be of research interest.

Table 1.3 identifies eight key techniques which, in combination, represent a powerful collection of methodologies that provide a wide range of analytical approaches. It is therefore obvious that any procedural undertakings that affect the validity of a single analytical technique have the potential to affect a broad range of methodologies. The 'critical points' in these key techniques must therefore be well characterised in order to minimise, counteract or, at the very least, understand their effect on the analytical data produced.

Table 1.3 *Key techniques and associated methodologies employed in analytical molecular biology*

Analytical technique	Method	Type of analysis*
DNA extraction	Various methodologies, usually a prerequisite for all the following analytical techniques	--
DNA quantification	Various methodologies, can be a prerequisite for all the following analytical techniques	--
Polymerase chain reaction (PCR)	Random amplified polymorphic DNA (RAPD)	Ch/Ql
	Amplified fragment length polymorphism (AFLP)	Ch/Ql
	Multiplex PCR, e.g. STR genotyping	Ch/D/Ql
	Nested PCR	Cf/D/Ql
	Quantitative PCR	M/Qt
	Cycle sequencing	Ch/Cf/Ql
	Cleaved amplified polymorphic sequence (CAPS)	D/Ch/Ql
Sequencing	Cycle sequencing	Ch/Cf/Ql/Qt
Hybridisation	Restriction fragment length polymorphisms (RFLP)	Ch/Ql
	Dot/slot blots	M/D/Qt/Ql
Restriction digests	AFLP, CAPS (PCR-related methods — see above)	Ch/Ql
	RFLP (hybridisation-related methods — see above)	Ch/Ql
Electrophoresis	PCR-related methods — see above	
	Single stranded conformational polymorphisms (SSCP)	Ch/Ql
	Pulse field gel electrophoresis (PFGE)	Ch/Ql
	Sequencing	Ch/Cf/Ql/Qt
Oligonucleotide synthesis	Prerequisite for all the PCR-related and some hybridisation methods -- see above	M/D/Qt/Ql
	Dot/slot blots	

*Refer to Table 1.1. D = detection, M = monitoring, Cf = confirmatory, Ch = characterisation, Ql = qualitative, Qt = quantitative.

1.7 Future Prospects and Considerations

The transfer time of a technique from the research laboratory to the analytical laboratory can vary. This could be dependent upon whether the new analysis is a further application of existing DNA technology, or whether it is an unfamiliar method using novel techniques and equipment. The former may require a shorter time period as reduced training, protocol preparation and validation could be required. In either case, a close working relationship between the researchers and analysts can ease the transition by building a clear understanding of each other's goals and requirements and working together on common ground.

Plans for the future of analytical applications appear to be progressing towards miniaturisation, parallelisation and automation.[3,4] In order to achieve this, improvements are required in the areas of sample preparation, assay technology, detection systems and data management. There is also a need to integrate the required steps in an economic way so that a given DNA analysis procedure can be performed substantially quicker and cheaper than existing tests.

Recent advances in the adoption of molecular biology, in particular PCR,[5] as an analytical tool continue to meet a wide demand for ever increasing improvement to levels of detection, accuracy, sensitivity and reliability. Quality should also be at the forefront of demands made on this evolving technology and this subject forms the core theme that runs throughout this book.

The acceptance of DNA profiling as an analytical tool has much to offer us as a lesson to be learnt. This innovative technology, first described by Jeffreys *et al.*,[6,7] was first used in a court of law in the 'Pitchfork' case in Lincolnshire in 1986. Since then, the validity of DNA data submitted as evidence in courts of law has been challenged. The stringent validation and quality processes that are now in place in today's forensic laboratories have therefore been, to some extent, driven by the pressures of the defence lawyers, continually challenging the analytical process both in this country and abroad. The presence of an equivalent pressure is not always evident in other areas of analytical molecular biology such as environmental or clinical testing. In these cases, the majority of the impetus for ensuring that appropriate data are produced as a matter of course lies with the professionalism of the analytical laboratory and the analysts involved. This is not a task to be undertaken light heartedly. It requires continual questioning and re-evaluation of the analytical approach, procedure, staff capabilities and applicability of the test. Analytical laboratories should, as a priority, work to maintain the confidence of the public and industrial customers by promoting the production of quality analytical data.

1.8 References and Further Reading

1. Saiki, R. K., Gelfand, D. H., Stoffe, S., Scharf, S. J., Higuchi, R., Horn, G. T., Mullis, K. B. and Erlich, H. A. 1988. Primer-directed enzymatic amplification of DNA with thermostable DNA polymerase. *Science* **239**: 487–491.

2. Saiki, R. K., Scharf, S., Faloona, F., Mullis, K. B., Horn, G. T., Erlich, H. A. and Arnheim, N. 1985. Enzymatic amplification of β-globin genomic sequences and restriction site analysis for diagnosis of sickle cell anemia. *Science* **230**: 1350–1354.
3. Allain, J.-P. 1995. Molecular diagnostics for infectious diseases: New approaches and applications. *Trends Biotechnol.* **13**: 413–415.
4. Abramowitz, S. 1996. Towards inexpensive DNA diagnostics. *Trends Biotechnol.* **14**: 397–400.
5. White, T. J. 1996. The future of PCR technology: diversification of technologies and applications. *Trends Biotechnol.* **14**: 478–483.
6. Jefferys, A. J., Wilson, S. L. and Thein, S. L. 1985. Hypervariable minisatellite regions in human DNA. *Nature* **314**: 67–73.
7. Jefferys, A. J., Wilson, S. L. and Thein, S. L. 1985. Individual-specific fingerprints of human DNA. *Nature* **316**: 76–79.

CHAPTER 2

Quality in the Analytical Molecular Biology Laboratory

GINNY C. SAUNDERS

2.1 Introduction

The need for valid practices to produce quality data cannot be disputed in any analytical environment; however, the route to consistently obtaining quality analytical data is not necessarily a clear and straightforward path. This chapter will attempt to introduce and highlight the many factors that can and do influence the ultimate goal of the analyst, namely getting the analysis right first time and every time.

There are four key criteria that must be met in order to obtain quality data:

1. A valid methodology
2. A quality assurance system
3. A quality control system
4. Trained and experienced analysts

It is important to note that, throughout the practical chapters of this manual, the assumption is made that the last three of these requirements have been fully met. The first, a valid methodology, is largely the subject of Section 2.4, with additional information presented in each practical chapter as appropriate.

Figure 2.1 demonstrates how these four criteria are inter-linked and work together to yield quality data. In turn, quality results are dependent on all of the criteria being met; a single criterion, even if met, cannot act effectively in isolation from the others. For example, trained and experienced analysts are required both to ensure that the chosen method is fit for its intended purpose and that it is correctly applied. To ensure that this is the case, the analyst will require a full understanding of the scientific principles of the methodology and its limitations. Equally, application of a valid methodology may be invalidated if equipment is not correctly calibrated, or samples are stored incorrectly. These are just two of the many factors that should be managed by a quality system. In addition, methodologies should stand up to external quality control through the

Valid methodology   Trained Analysts

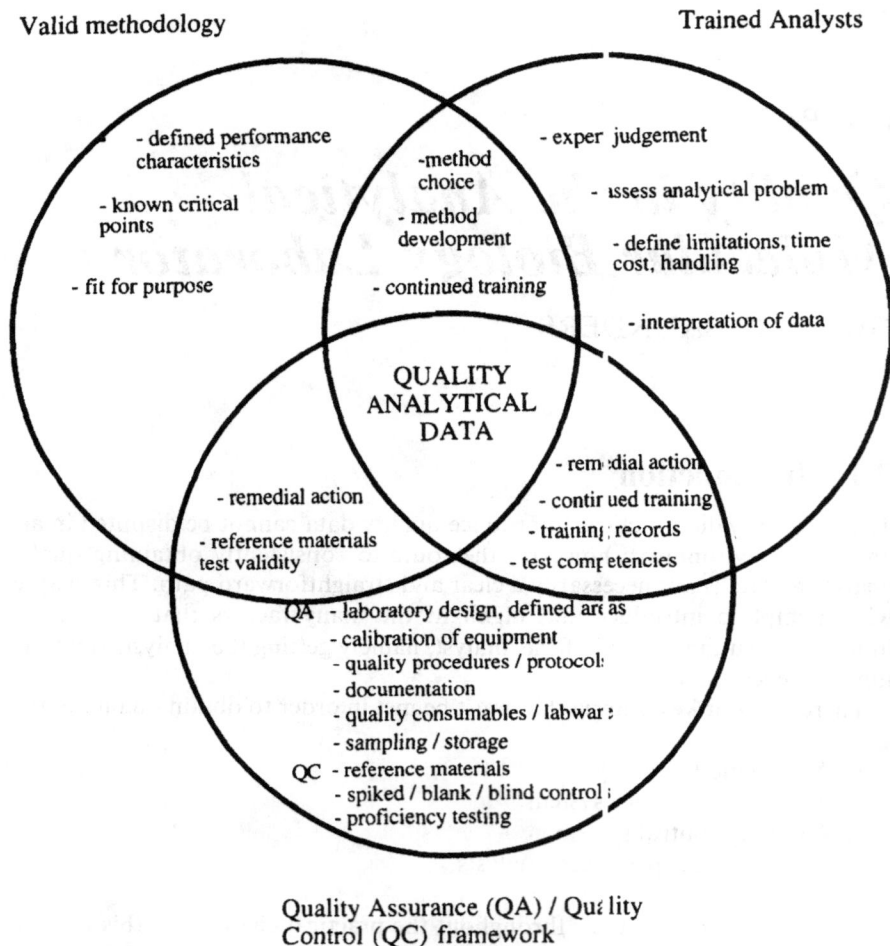

- defined performance characteristics
- known critical points
- fit for purpose

-method choice
- method development
- continued training

- exper judgement
- assess analytical problem
- define limitations, time cost, handling
- interpretation of data

QUALITY ANALYTICAL DATA

- remedial action
- reference materials test validity

- remedial action
- contir ued training
- trainin records
- test comp etencies

QA - laboratory design, defined ar as
- calibration of equipment
- quality procedures / protocol
- documentation
- quality consumables / labwar
- sampling / storage
QC - reference materials
- spiked / blank / blind control
- proficiency testing

Quality Assurance (QA) / Qua lity Control (QC) framework

Figure 2.1 *Schematic diagram showing the criteria that must be met in order to obtain quality data*

use of reference standards or by comparing analyses with those undertaken elsewhere. Quality assurance schemes and experienced analysts should play an important part in determining the cause of any disparity between expected and actual results and defining any remedial or preventative action required.

Adoption of a uniform understanding of terminology commonly used in this area is also required to ensure that these criteria are consistently interpreted and fully met. This chapter sets out to introduce and define various quality and validation terms. Some general terms used throughout the manual are given in Table 2.1.

The Valid Analytical Measurement (VAM) Initiative aims to present a holistic approach to obtaining quality data by presenting six principles for a

Table 2.1 *Definition of relevant terms*

Term	Definition
Quality	Quality in an analytical environment ensures the provision of data that meets the needs of the enquiry or customer, instils confidence in the receiver of the data and those who generated it, and represents value for money[1]
Validation	Validation of a methodology establishes, by systematic laboratory studies, that the method is fit for purpose, i.e. its performance characteristics (Section 2.4.1) are capable of producing results in line with the needs of the analytical problem[2]
Technique	Scientific principle (PCR, hybridisation)[3]
Method	Distinct application of a technique (multiplex PCR, Southern blotting)[3]
Procedure	Written directions to perform a method[3]
Protocol	Set of definitive directions that must be followed, if the resulting data are to be considered fit for their intended purpose[3]

laboratory to follow. These principles can be applied in a self-regulatory manner and in no way interfere or conflict with the remit of a quality system.

2.2 VAM Principles

The role of the VAM Initiative supported by the Department of Trade and Industry is to promote the production of quality data. It does so by establishing six principles that are generic, fundamental truths of good analytical practice, irrespective of the scientific discipline to which they are applied.[4] These principles, set out below, are in no way intended to replace, contradict or diminish the quality assurance, quality control or accreditation schemes that may be in place. In fact, a positive requirement for such schemes is supported by the principles.

1. *Analytical measurements should be made to satisfy an agreed requirement.* Analysis is generally undertaken to answer a specific question or to provide a solution to a problem. This principle promotes the use of the most suitable analytical approach in order to ensure that the question will be directly addressed by the data produced. This is dealt with in Section 2.3, where the analytical context is discussed.
2. *Analytical measurements should be made using methods and equipment which have been tested to ensure they are fit for their purpose.* Guidelines on how to validate a method and ensure that it is fit for its intended use are given in Section 2.4. Equipment calibration is usually addressed by a quality assurance scheme; this is dealt with in Section 2.5.
3. *Staff making analytical measurements should be both qualified and competent to undertake the task.* Quality assurance and quality control schemes

should address the continued training and asse ssment of staff (see Section 2.5).

4. *There should be a regular independent assessm nt of the technical performance of a laboratory.* External assessments of performance are useful to confirm that procedures are being followed. Proficiency testing schemes (see Section 2.6) offer the opportunity to compare in-house results to analyses carried out in other laboratories.

5. *Analytical measurements made in one location should be consistent with those elsewhere.* A valid methodology can be tested on samples of known consistency, such as reference materials (RMs) if available, or spiked samples (see Section 2.7). These should be employed on a regular basis.

6. *Organisations making analytical measurements should have well defined quality control (QC) and quality assurance (QA) procedures.* QA schemes can assist in preventing errors by ensuring that trained staff, calibrated equipment, quality protocols and valid methodologies are used. QC confirms the quality of data obtained by the use of control samples and helps confirm that the QA scheme is operating correctly (see Section 2.5).

Where the act of taking measurements is referred to, this should not solely be viewed as the production of quantitative data but can equally be interpreted as a basis for determining qualitative information. Examples of qualitative measurements include identification and characterisation techniques such as DNA profiling, whether in the form of RAPD, STR or RFLP generated data, or simple screening, positive/negative-type assays. Such qualitative data can also be expressed quantitatively (e.g. the sizing of DNA bands on a gel) and this too requires validation. As much of the analysis current y undertaken in analytical molecular biology is of a qualitative nature, this is an important detail. No lesser emphasis is placed on the need for valid qualitative data as valid quantitative data. The tools to measure the validity or quality of the data may differ, however, and this is discussed in Section 2.4.

Whilst it is clear that the VAM approach to quality requires support and implementation at management level, the ethos of these principles needs to be understood and supported by all levels of staff involved in analysis.

2.3 Analytical Context

Analysis is carried out to address a specific question or enquiry and, on the basis of the data obtained, a decision making or solution forming process generally follows. For example, human STR analysis may assist in determining whether an incriminating bloodstain could have originated from a particular suspect; genetic analysis of a patient may determine medical treatment or influence lifestyle options; detection of a pathogen may determine whether it is present at acceptable levels or if preventative measures should be taken; initial analysis may ascertain whether further, more extensive, confirmatory analysis is required. The influence of analytical data obtained can therefore be far reaching and could have severe consequences if it cannot be fully relied upon.

It is therefore vital that analysis is carried out to the highest possible standard and that a structured analytical approach is followed. For efficient analysis, careful pre-analysis planning of the work can save valuable time later on. A flawed approach may produce experimentally valid data that do not directly address the enquiry and so are not valid in that particular situation. Incorrect sample collection or storage could produce erratic results even when a valid method is applied. The importance of pre-analytical planning cannot be over stressed.

A broad outline of a possible sequence of individual tasks involved in analysis is outlined below and summarised in Figure 2.2.

1. *Define the analytical enquiry.* A clear understanding of what is required of the analytical data is vital to applying the correct technical and methodological approach. Verify whether qualitative, semi-quantitative or quantitative end-point analysis is required. Define the criticality of the data and the performance levels (see Section 2.4) required of a method.

2. *Assess the sample.* Confirm the nature of the sample (matrix, size) and form a strategy for sample collection, transportation, storage and preparation. Be as familiar as possible with the history of the sample, availability of sample, number or throughput of samples and the controls available or required (further discussed in Section 3.3). Ensure a working sample identification system is in place so that each sample has a unique identifier.

3. *Establish constraints.* Establish constraints placed on the analysis with respect to available time and resources (analysts, equipment and cost) and any safety considerations. This stage could determine whether the undertaking of the work is actually viable.

4. *Define the technical approach required.* Decide upon a suitable technique, or a combination of techniques, that best meets the needs of the analytical enquiry. Consider the scope of information the data from the technique could or could not yield, with respect to both method performance (e.g. level of sensitivity or precision required) and interpretation of data (e.g. a negative result could imply that a microorganism may be present at concentrations below the detection limits; equally, a positive result could imply that dead cells of the target microorganism were detected). A level of confidence may also need to be considered, e.g. a provision for the accountability of false negatives and false positives, or definition of the accuracy of quantitative analyses.

5. *Select and/or develop methodologies.* Based on the information gathered in points 1–4 above, select a compatible methodology either from published procedures, either by building a procedure from other methods currently in use or by developing a new methodology. Selection or development will be based on known or theorised characteristics of how the method performs and should be expected to generate data that meet the needs of the enquiry and the sample in question. For example, it may be unnecessary to employ a very sensitive method for the detection of an analyte

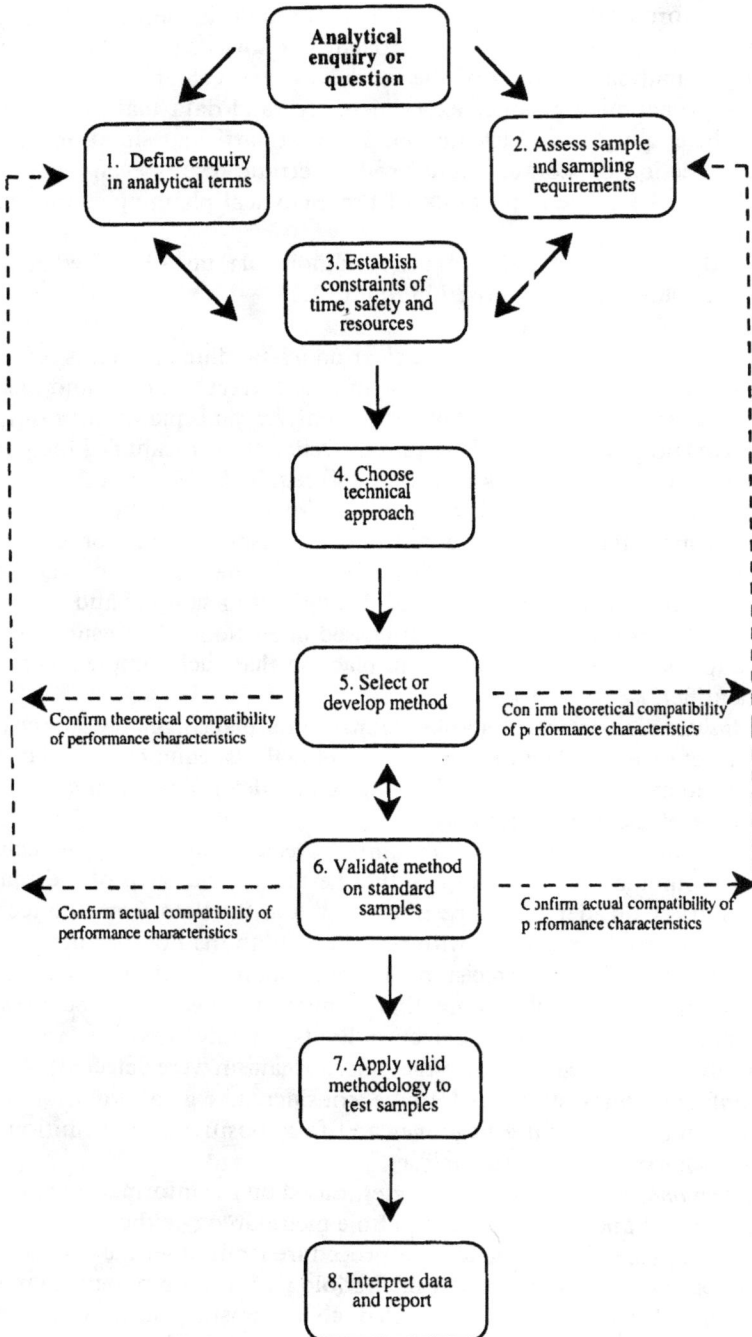

Figure 2.2 *Schematic diagram to illustrate the analytical process*

occurring at high concentrations. Ease of use, cost, hands on and total analytical time required may also be considerations. Prepare a draft protocol.

6. *Validate method.* Identify critical parameters that affect the robustness of the methodology and define pertinent performance characteristics (see Section 2.4) to be validated. Investigate performance characteristics and critical parameters by practical examination using relevant reference materials or spiked materials. Revise the protocol as required into its final form. Even if a method has published performance characteristics, these will still have to be confirmed or re-evaluated in the test laboratory, usually by future analysts and using the equipment that will finally be used. Criticality of data and uniqueness of sample may determine the extent of validation required. If, at this stage, the performance of the chosen method does not meet the requirements of the analysis as defined in 1, 2 and 3, then a re-examination of alternative analytical approaches should be undertaken or the requirements redefined in consultation with the customer. The validation process should be well documented and should be available to all operators subsequently using the methodology.

7. *Apply the validated methodology.* Analyse samples and appropriate quality control samples (see Section 2.5).

8. *Interpret the data and report.* Data should be presented and interpreted to encompass any caveats or limitations defined by examination of the performance characteristics during the validation process.

This approach is not necessarily a one-way system, as shown in Figure 2.2. In particular, the process of method development followed by validation may not necessarily be sequential. More realistically, there is not always an obvious point where development ends and validation starts, and these processes are likely to meld into one another. If a suitable method cannot be found that meets both the requirement of the analysis and the sample, the expectation of the analytical information as determined in point 1 may have to be redefined. Redefinition can only take place after close consultation with the source of the enquiry, i.e. the customer.

2.4 Method Validation

Validation is the practical process of determining the suitability of a methodology for providing useful analytical data. Method validation should be undertaken by analysts who are broadly competent with the techniques and using suitable and calibrated instrumentation.

It should be noted that a valid method is not necessarily free from error. Bias, a type of error, can be acceptable if empirically determined. The validation process, however, does define a range of performance characteristics within which it has been confirmed that the method can yield acceptable results. The use of a valid method is therefore limited by its application. The data obtained must lie within the boundary of the performance characteristics, as assessed, which deemed them fit for an intended use. For example, if the range in sample size for which the method has been deemed to produce reproducible results is 1–10 g, then the method is only considered to be valid when applied to samples within that range. Equally, if the sensitivity of a method was empirically determined to be a detection limit of 10% adulterant in a matrix, it cannot be relied upon to quantitatively detect adulteration of less than 10%.

If a method is changed in any way, for example a different chemical or enzyme is used, new equipment is purchased or if the method is applied to a different matrix, the validation process may have to be re-examined. In addition to the performance characteristics set out below, different pieces of equipment, laboratory temperature and other common variables may also need to be incorporated into method validation or duly noted.

2.4.1 Measures of Validity

Performance characteristics (sometimes referred to as assay parameters in clinical diagnostics) give definition to how a method functions under specified conditions. Factors that affect the performance characteristics of a method or technique and therefore its validity are collectively referred to as *critical points* throughout the course of this manual.

Knowledge of performance allows the analyst to decide whether a method is valid for the production of data in a given application. Typical performance characteristics such as accuracy, bias, precision, range, recovery, repeatability, reproducibility, sensitivity and selectivity are set out in this section. Their relative import in a given situation will depend on the intended final use of the analytical data. It is important to note that some of the terms below have different meanings when applied to different analytical disciplines. Official definitions and terminology can be obtained from the IUPAC Compendium of Chemical Technology[5] and ISO 3534–1.[6] Terms and their explanation used throughout this manual are given below.

Accuracy — A measure of the difference between the true value and the mean of a set of analytically determined values (see Figure 2.3). Determination of the true value may be problematic, however, and the accuracy of that measurement may not be calculable! Ideally, the true value is determined by the analysis of appropriate reference materials (RMs) or standards (see Section 2.6) where available, otherwise through the use of spiked samples. The use of RMs or standards are greatly preferred as comparisons between laboratories are then possible.

Bias — Error introduced to the analysis systematically; one or more factors may contribute to bias, such as sample storage effects or improper calibration of equipment. The bias of a result may therefore be constant or at least predictive.

Precision — Comparison of repetitive quantitative measures for the whole method, not just the quantification step. A method is considered precise if there is little or no variation between the measurements. Precision should be determined both with respect to repeatability and reproducibility (see below) and for the measurement of an analyte at each relevant concentration. It is

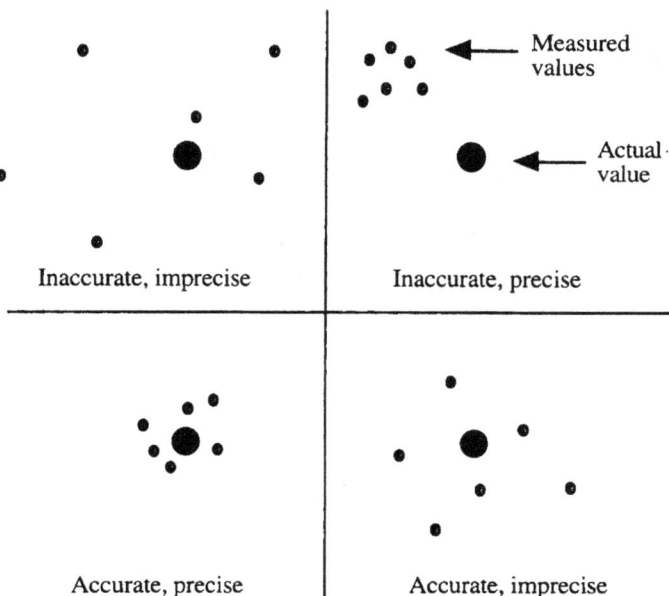

Figure 2.3 *Schematic illustration of accuracy and precision, see text for definitions*

important to note that the precision of the data obtained is not a reflection of its accuracy (see Figure 2.3); therefore, where possible, accuracy must also be determined. Precision can be measured as:

- standard deviation under repeatable conditions
- standard deviation under reproducible conditions

Range — The interval between the upper and the lower concentration of an analyte in the sample for which it has been determined that the method is suitable. This could also be expressed as the mass or volume of matrix for which it has been determined that the method is suitable. Analytical techniques cannot necessarily be scaled up or down indiscriminately.

Recovery — Fraction or percentage of actual amount of analyte obtained after one or more stages of analysis. This can be determined by employing RMs, where possible, or spiked materials (see Section 2.6).

Repeatability — Quantitative or qualitative measurements made using a given method on a particular sample under similar conditions, i.e. same operator, same laboratory, over a narrow time period. Repeatability could be expressed as a percentage of analyses that produced identical qualitative results out of the total number of analyses undertaken, or as the standard deviation of quantitative data (see Precision).

Reproducibility — Quantitative or qualitative measurements made using a given method on a particular sample under different conditions, i.e. different operator, different laboratory, over a long time pe iod. Reproducibility could be expressed as a percentage of analyses that prc duced identical qualitative results out of the total number of analyses unde taken, or as the standard deviation of quantitative data (see Precision).

Robustness (ruggedness) — A measure of a mc thod's capacity to remain unaffected by small variations in the main parametc rs. Introduction of deliberate variations in the validation process will defint critical parameters in the method; for example, incubation times and temp :ratures, concentrations of buffer constituents, pH, amount of components .idded, degree and type of mixing. An assessment of robustness can influence he production of protocols to ensure that correct emphasis is given to critical an l non-critical parameters or tasks.

Sensitivity (limit of detection/detectability) — Smallest concentration of target analyte that can be determined and is disting uishable from a zero result. For quantitative determination of zero, a zero calil rator (a sample known *not* to contain the analyte of interest) is analysed 1)–20 times, the mean and standard deviation of the data obtained are calcula ed and the mean signal ± 2 standard deviations is considered the analytical sensitivity.[7,8] It should be noted that the biological definition of sensitivity given here varies from the ISO chemical definition. Sensitivity could be expressed a s:

- number of cells per mass of matrix detectablc
- % of adulterant in a matrix detectable
- mass of DNA required for qualitative analysis
- mass of DNA detectable at end point analysi;
- copies of a gene or genome per volume or miss detectable

Sensitivity must be assessed at all detection levels and for a range of matrix masses as it does not necessarily follow that if 100 microbial cells can be detected in 1 g of soil by PCR, then 10 cells will be detectablc in 0.1 g of soil.

Sensitivity can also be expressed as the ability of a reaction to detect an analyte without the production of false negatives.[7,9] This use of the term is often applied in clinical diagnostic environments. Sensitivity could be expressed in percentage terms as a detectability index (DI). Fcr example, if 100 samples known to contain a certain analyte, such as a genc mutation, were processed, and of those 100 samples 95 tested positive and 5 tes ed negative (false negative), this would produce a DI of 95%:

$$\frac{\text{true positives (95)}}{\text{true positives (95)} + \text{false negatives (5)}} = \text{DI of positives} = 95\%$$

Selectivity (discriminability, specificity) — The ability to distinguish the target analyte from other analytes. Poor specificity of a reaction indicates that other substances can interfere with the analysis. Selectivity of a reaction can change as one of the parameters is changed. Fc r example, a PCR at the optimised annealing temperature will produce a s ngle, specific amplification product, but if the annealing temperature is changed and the reaction made less stringent, multiple non-specific products can be generated.

Selectivity is another term applied in a slightly different manner in the field of clinical diagnostics.[7,9] Here it is viewed as the ability of an assay to give consistently a negative result for known negatives. This is expressed in percentage terms as a detectability index. See the example given under sensitivity:

$$\frac{\text{true negative}}{\text{true negative} + \text{false positive}} = \text{DI of negatives}$$

Specificity can be considered as a special case of selectivity, i.e. a process is specific if it is 100% selective.

2.4.2 Valid or Not Valid?

It is evident that method validation could be an extensive and time consuming task. In reality a limited amount of time and resources are usually allocated to the task. Therefore the scope of the validation investigation and the point at which a method is considered valid has to be decided upon. The degree of validation required is dependent on many factors; however, as general guide, validation can be influenced by four main criteria:

- the criticality of the data
- the uniqueness of the sample
- the robustness of the technique
- the expected utilisation of the technique

For example, a scene of crime sample would be considered unique and the data derived from it highly critical; therefore maximum validation may be required. Screening of plants for the detection of a genetic modification might suggest high availability of sample and low criticality as confirmatory tests could follow; therefore minimum validation may be required. A sample may be considered unique both with respect to time (it may represent a unique moment in the progression of an infection), quantity (single hair at scene of crime) or difficulty of sampling (biopsy material). If a technique is to be applied to a limited number of ad hoc samples and applied over a short time period, it may not be appropriate to carry out extensive validation that might be associated with the evaluation of a technique for long-term routine laboratory use. For example, inter-operator and long-term time variables may not necessarily be an issue. In most cases it is perhaps usual for a compromise to be struck between resource restraints (time, costs and current expertise) and the need to validate a methodology. This matter ultimately comes down to professional judgement. It should be emphasised that 'no validation' is not an option. All analytical data should be considered as critical to a certain degree from the customer's perspective. Furthermore, a plentiful supply of sample material does not mean that it is acceptable to employ invalid methodology; it remains a fundamental aim of the analyst to get the analysis right first time and every time and provide quality data and value for money to the customer.

A broad outline procedure for validation may be useful in a laboratory. The example given below applies to a single technique. An analytical process may consist of many different techniques, all of which require validation.

1. Write draft protocol.
2. Define criticality of data.
3. Define robustness of technique. Sometimes this may have occurred during method development and critical points in the technique may already be known. If this is not the case, it may be prudent to commence with a study of robustness. A useful by-product of the robustness test is the identification of critical points in the method which can indicate where the emphasis of the validation study should be placed. A statistically structured approach to testing robustness (ruggedness) can be undertaken, as discussed by Youden and Steiner.[10] Amend protocol to give emphasis to critical parameters.
4. Identify performance characteristics applicable to the method and analytical task. Typically, a quantitative analysis may require validation of sensitivity, precision, accuracy, bias, specificity, recovery, range and robustness.
5. Identify the performance characteristics that require validation and their most logical order of investigation, i.e. specificity > sensitivity > range > recovery > precision > bias > accuracy. The most logical order will vary, depending on the technique, analytical requirements and sample.
6. Assess performance characteristics of methodology using a consistent supply of well characterised, reference, standard or spiked material.
7. Document performance characteristic investigation.
8. Assess whether the data obtained deems the method valid or fit for its intended application, e.g.
   - the technique may need to be robust enough to be unaffected by such small changes as incubation temperature changes of $\pm 1\,°C$ or incubation times of ± 10 min, etc.
   - selectivity may need to have a DI greater than 95% if the technique is used to discriminate closely related targets
   - a sensitivity level minimum of a 5% adulterant detection may be required as this is the legal requirement
   - the technique should be suited to sample sizes of 1–10 g
9. Define limitations of the methodology.
10. Write final detailed protocol. Examples of protocols for technical procedures carried out by the FBI are given by Easteal *et al.*[11]

A summary of the validation data should be available with the protocol; this will clearly allow users to decide whether it is suited to a given application, and identify any critical parameters in the method.

2.5 Quality Assurance

It is of paramount importance to an analytical laboratory that a quality framework is set up in order to ensure that all analysis is performed with the highest possible assurance of quality. Quality systems are relatively new to analysis and are generally formed to monitor the performance of day-to-day operations.

- *Quality assurance* — A QA system defines, and sets out through documentation, procedures that need to be continually adhered to in order to obtain quality data persistently. This typically covers the production of quality protocols (contained in a quality manual), monitoring of the laboratory environment, calibration of equipment, employment of suitable staff and their training, labelling and content control (for chemicals and samples), standard of consumables, use of valid methodologies, proficiency testing and quality control. It also encompasses a monitoring or auditing system to ensure that all documentation and procedures are completed as required.
- *Quality control* — A QC system employs control samples to ensure that a technique is operating correctly and confirms data quality. Control samples are processed alongside samples to be analysed and comprise of well-characterised, blind, blank and duplicate samples. Any deviation from expected results will require corrective action and the co-analysed batch of samples will require re-analysis. As a guide, QC samples should make up approximately 5% of the samples analysed; this could be decreased to 1–2% for robust high-throughput types of analyses or increased to 50% when carrying out complex *ad hoc* forensic analyses. Routine quality control is usually assessed inside an organisation. External quality control is available in the form of proficiency testing schemes (Section 2.5.4). Control charts can be used routinely to record data obtained from control analyses. These can highlight differences between operators, signal degradation of consumables or suggest that equipment calibration is required.

2.5.1 Calibration

Equipment used for analytical measurements should be fully calibrated using calibration standards or certified reference materials and used within its limitations of accuracy or capacity according to the manufacturer's instructions. Typical equipment requiring calibration in a molecular biology laboratory includes:

- balances
- pipettes
- timers
- waterbaths

- thermometers
- spectrophotometers, including UV and fluorometric
- pH meters
- autoclaves
- refrigerators and freezers
- incubators, ovens and heating blocks
- thermal cyclers

2.5.2 Contamination

Contamination can be a major source of poor quality data in a molecular biology laboratory. This can largely be avoided by sensible laboratory practice. For example, protective clothing should be worn by all occupants of the laboratory, small aliquots of reagents should be used to avoid extended exposure, samples should remain sealed at all times when not in use, the production of aerosols or fine, particulate biological material should be avoided or carried out in a suitably contained space, and all equipment, utensils, materials and benches should be regularly cleaned or sterilised as required. In laboratories carrying out analysis of human DNA, samples from analysts undertaking the work should be characterised as, in some cases, this may identify if contamination is caused by a member of the laboratory.

The layout of the laboratory may require areas designated to a certain task in order to avoid cross-contamination. Typically, a laboratory carrying out PCR will have a one-way flow system for sample processing. A pre-PCR DNA extraction area should be physically separate from the DNA-free PCR set-up room and a further post-PCR detection area. Dedicated equipment and protective clothing should be freely available in each location, thereby minimising contamination by the movement of materials between areas (see Section 5.3.4 for further details).

2.5.3 Types of Quality Assurance Standards and Accreditation Schemes

A laboratory can implement its own QA system or perhaps more commonly choose to comply with an established QA standard.[12-15] Accreditation is defined as 'formal recognition that a laboratory is competent to carry out specific tasks'.[16] Accreditation is obtained through assessment by an independent audit to determine whether there is compliance to a recognised QA scheme. Therefore each formal QA scheme has its corresponding compliance or accreditation scheme. The four schemes perhaps most relevant to analytical molecular biology are discussed and summarised in Table 2.2.

2.5.3.1 *ISO 9000 Series of Quality System Standards*

The ISO standards represent 'distinct forms of quality system requirements suitable for the purpose of a supplier (laboratory) wishing to demonstrate its

Table 2.2 *Standards and accreditation schemes available*

International standard	UK standard	Accreditation scheme
ISO 9001/2/3	BS-EN-ISO 9001/2/3	National Accreditation of Certification Bodies (NACB) accredit certification bodies as competent to assess organisations against ISO 9000 and confer certification of compliance.
OECD-GLP	Dept. of Health (DoH), Environmental Protection Agency (EPA), Food and Drug Administration (FDA) Principles	GLP Monitoring Authority register organisations' studies as being GLP compliant.
ISO Guide 25 EN 45001	NAMAS M10	United Kingdom Accreditation Service (UKAS)
CPA standard	CPA standard	Clinical Pathology Accreditation (UK) Ltd

capability of the supplier by external parties'. Generally, ISO 9001 is the ISO standard applied to laboratory services engaged in design and development of analytical procedures and incorporates the previous BS 5750 quality standards. It should be noted, however, that the ISO standards apply to quality management only, thereby ensuring that set procedures are in place and are being followed; it does not make any assessment of technical competence.

2.5.3.2 Organisation of Economic Co-operation and Development — Good Laboratory Practice

The GLP Principles, published in the UK by the Department of Health, are primarily a management tool applied to toxicological studies conducted in support of production registration on, for example, pharmaceuticals, agrochemicals, cosmetics and food additives.

2.5.3.3 National Accreditation of Measurement and Sampling

NAMAS accreditation is applicable to objective measurements carried out for calibration or testing purposes in a variety of sectors and is covered by a single standard, M10. In general, the standard sets out the inputs required to support a quality system and NAMAS accredits laboratories to carry out specific tests. The quality of the data produced is usually measured by contributing to proficiency testing (Section 2.5.4) or other external control schemes.

2.5.3.4   Clinical Pathology Accreditation Standards

Standards are intended to meet most medical diagnostic laboratory services, including clinical biochemistry, microbiology, haematology, virology, immunology and molecular genetics. More information is given in their 'Accreditation Handbook'.

2.5.4   Proficiency Testing Schemes

PT schemes, sometimes called External Quality Assessment (EQA) schemes in the clinical sector, can be run on a co-operative basis between laboratories carrying out comparable analyses or by a single organisation. The scheme may or may not be commercially based. A test sample is distributed to participating laboratories for analysis and the resulting data are then collated and analysed by the scheme organiser. These schemes allow laboratories to compare their results with those obtained elsewhere or the true measurement, when available. This can lead to a general improvement in analytical technique through the identification of errors. PT schemes can also be a useful way of comparing the merits of different analytical approaches. The value of PT is recognised by quality standards such as those of UKAS, GLP and ISO 9000.

2.6   Reference Materials

A reference material (RM) is defined in the ISO guide 30[17] as a material or substance, one or more properties of which are sufficiently well established to be used for:

- calibration of an apparatus
- assessment of a measurement method
- assigning values to materials

Once the property value(s) (e.g. trace element, fat or metal content of a given matrix) of a particular RM have been established by measurement, they are in effect 'stored' by the reference material until its expiry date. The values are then transferred when the RM itself is conveyed from one site to another. RMs are therefore valuable resources used to underpin consistency and comparability of analyses.

In order for a RM to perform its role, the following technical criteria must be met:

1. The RM itself and the property value(s) embodied in it should be stable for an acceptable time-span, under realistic conditions of storage, transport and use.
2. The RM should be sufficiently homogenous that the property value(s) measured on one portion of the batch should apply to any other portion of the batch within acceptable limits of uncertainty; in cases of heterogeneity

of the large batch, it may be necessary to certify each unit from the batch separately.

3. The property value(s) of the RM should have been established with a precision and an accuracy sufficient to the end use(s) of the RM. Accurate is defined as 'made by a method having negligible systematic error or bias and by means of measuring instruments or material measures which are traceable to national measurement standards'.
4. Clear documentation concerning the RM and its established property values should be available.

Certified reference materials (CRMs) are also sometimes referred to, and are defined as 'a reference material, one or more of whose property values are certified by a technically valid procedure, accompanied by or traceable to a certificate or other documentation which is issued by a certifying body'.

The stringent requirements of a reference material do not complement the complex molecular composition of biological materials, which can be inherently heterogeneous, lack stability and difficult to quantify. However, numerous RMs have properties which, because they cannot be completely characterised by physical and chemical methods alone, cannot be measured in mass or amount of substance units. This is perfectly acceptable. Such reference materials include current biological RMs developed by the World Health Organisation (WHO). These are available through the National Institute for Biological Standards and Control (NIBSC) and include a range of vaccines, hormones, antibiotics and other biological substances of interest to the pharmaceutical industry and clinical research and development. The National Institute of Standards and Technology (NIST) in the USA also produce DNA-based reference materials for the quality assurance of forensic and paternity procedures (RFLP and PCR based profiling), available in the UK through Promochem Ltd.

The recent arrival of molecular biology to the analytical arena means that the current range of RMs available from the Office of Reference Materials or indeed any other source is, at present, noticeable for its limited availability of biological materials. Therefore it is unfortunate that there are many measurements in analytical molecular biology for which an appropriate reference material does not exist. RMs that would prove useful in analytical molecular biology include:

1. *An internal PCR control* — A well characterised and highly specific control PCR reaction that could be included in all PCRs. Use of such a control could assess operator, reagent and equipment performance. False negative reactions, due to the presence of inhibitors, could also be detected by a failure to amplify the control.
2. *Reference food, medical or environmental matrices* — Well characterised matrices with respect to species identification, inhibitors present or genome characterisation (known DNA profiles). These could assist in the accurate quantification of biological adulteration, design of DNA extraction methods or optimisation of profiling techniques.
3. *Nucleic acid quantification standards* — Stable standards containing

known concentrations of nucleic acids (single stranded, double stranded, high molecular weight, degraded, etc.) could assist in assessing the accuracy and reproducibility of quantification measurements.

2.6.1 Alternatives to Reference Materials

Frequently used alternatives to reference materials include spiked matrices and in-house or commercial standards (so-called as they have not necessarily undergone the rigorous tests, characterisations and documentation required of a reference material).

A variety of biological matrices (such as soil, water, foodstuffs and biological fluids) can be spiked with target cells or DNA, or admixtures of matrices can be used to mimic real situations. Disadvantages of this approach include the difficulty of reliably preparing homogenous samples and batch-to-batch variation. During the process of spiking, where cells are invariably taken from a nutrient-rich culture and added directly to the biological matrix, cells may not react predictably in their new environment. A considerable percentage of the cells may lyse within a short time period as it is difficult to simulate actual matrix conditions such as pH, temperature and oxygen content. In addition, spiked samples may have different physical properties to those of real samples, where the cells or nucleic acids have been exposed to the matrix for a much greater period of time. Such properties might include different target adhesion strengths or limited access to naturally occurring matrix cavities. If, however, the spiked cells are incubated with a matrix for a period of time, enrichment of the cells may occur and the initial quantification of cell material will be lost.

In-house standards (such as extracted DNA, spiked materials or admixtures) are a useful alternative, although as they are not shared with comparable laboratories, reproducibility of a qualitative or quantitative nature, or quantitative bias, cannot normally be assessed. Traditionally, in-house standards are calibrated against RMs (when available), cutting down on cost of RMs and extending the usage of a single supply.

Typical commercial standards that are currently employed in analytical molecular biology include DNA and protein molecular size markers and allelic ladders for genotyping applications. Also available are highly pure DNA/RNA samples and numerous positive control components supplied with a vast array of detection or analysis kits.

2.7 Summary

Quality is not always quantifiable but a general rule of thumb assessment decrees that 'you only get out what you put in', a true commitment to the provision of quality is therefore required from both organisations and individuals alike. It should also be stressed that quality analytical data can only be consistently obtained if all the required criteria (valid methodology, QA/QC frameworks and competent staff) are in place and fully operational. This chapter has only touched on some of the complex issues surrounding the quality of analytical data and further reading is therefore recommended before implementation of any of the processes covered here is considered. Some useful addresses are given in Section 2.8.

2.8 Further Reading and Information

Reference materials
Office of Reference Materials. Laboratory of the Government Chemist, Queen's Road, Teddington, Middlesex TW11 0LY, UK
National Institute for Biological Standards and Controls, Blanche Lane, South Mimms, Potters Bar, Herts. EN6 3QG, UK
National Institute of Standards & Technology, Gaithersburg, MD 20899, USA
Promochem Ltd., 6S Mundells, Welwyn Garden City, Herts. AL7 1EP, UK

Proficiency testing
Proficiency Testing Advisory Service, Laboratory of the Government Chemist, Queen's Road, Teddington, Middlesex TW11 0LY, UK (Fax +44 (0)181 943 2767). Will supply information from the European Proficiency Testing Scheme Database

Accreditation schemes
Campden Laboratory Accreditation Scheme (CLAS), Campden and Chorleywood Food Research Association, Chipping Campden, Gloucester GL55 6LD, UK (applicable to chemistry, microbiology, and physical testing in the food, drink and allied industries)
Clinical Pathology Accreditation (CPA), 45 Rutland Park, Botanical Gardens, Sheffield S10 2PB, UK
European Accreditation of Certification Bodies (EAC, NACB, UKAS), Audley House, 13 Palace Street, London SW1E 5HS, UK
International Accreditation Forum (IAF), American National Standards Institute, 655 15th Street NW, Suite 300, Washington, DC 20005–5794, USA
LABCRED Laboratory Accreditation Scheme, Law Laboratories Ltd., Blakelands House, 400 Aldridge Road, Great Barr, Birmingham B44 8BH, UK (accreditation of chemical and microbiological testing services in the food industry)
National Accreditation of Measurement and Sampling (NAMAS), UK Accreditation Service (UKAS), Queen's Road, Teddington, Middlesex TW11 0NA, UK
UK GLP Monitoring Authority, Department of Health, Room 1801, Market Towers, 1 Nine Elms Lane, London SW8 5NQ, UK

2.9 References

1. Prichard, E. 1995. Quality in the Analytical Chemistry Laboratory. Wiley, Chichester.

 2. International Guide to Quality in Analytical Chemistry. An Aid to Accreditation. CITAC Guide 1. Dec 1995.
 3. Taylor, J. K. 1983. Validation of analytical methods. *Anal. Chem.* **55**: 600A–608A.
 4. The Manager's Guide to VAM. September 1996. Produced by LGC and NPL on behalf of the DTI.
 5. IUPAC Compendium of Chemical Technology. 1985. Blackwell, Oxford.
 6. ISO 3534–1.
 7. Davies, C. 1994. Concepts. In: The Immunoassay Handbook (ed. Wild, D.), pp. 83–116. Stockton Press, New York.
 8. Smith, M. 1994. Method evaluation. In: The Immunoassay Handbook (ed. Wild, D.), pp. 256–262. Stockton Press, New York.
 9. Pappas, M. G. 1994. Biotechnology product development. In: The Biotech Business Handbook, pp, 134–139. Humana Press, Totowa, NJ.
10. Youden, W. J. and Steiner, E. H. 1975. Statistical Manual of the Association of Official Analytical Chemists. AOAC, Washington.
11. Easteal, S., McLeod, N. and Reed, K. 1991. Appendix C. In: DNA Profiling. Principles, Pitfalls and Potential. Harwood Academic Publishers, Switzerland.
12. Accreditation for chemical laboratories: guidance on the interpretation of the EN45000 series of standards and ISO/IEC guide 25. EURACHEM Guidance Document No. 1/WELAC Guidance Document No. WGD 2, April 1993.
13. Burnett, D. 1996. Understanding Accreditation in Laboratory Medicine. ACB Venture Publications, Cambridge, UK.
14. Devereux, I. 1996. Implementing NAMAS (now UKAS) in a laboratory. QWTS. March 12–19.
15. Stenhouse, S. A. R. and Middleton-Price, H. 1996. Quality assurance in molecular diagnosis: the UK experience. In: Molecular Diagnosis of Genetic Diseases (ed. Elles, R.), pp. 341–352. Humana Press, Totowa, NJ.
16. NAMAS M10. 1989.
17. ISO Guide 30

CHAPTER 3
DNA Extraction

GINNY C. SAUNDERS

3.1 Introduction

The isolation of genomic DNA is, in many cases, a fundamental requirement for analytical molecular biological procedures. This chapter is concerned with situations where DNA extraction forms the first step in the analytical process. Inevitably, in such situations, without the successful execution of this step, further analysis such as detection and identification techniques (including RAPD, STR, sequencing and PCR detection of specific genes or mutations) would not be possible.

In cases where a sample is in limited supply (such as archival material) or even unique (such as scene of crime forensic samples), successful extraction of the valuable DNA analyte is of paramount importance. Equally, even if the sample is in plentiful supply, a well characterised and robust DNA extraction technique is required that will reproducibly isolate DNA within well defined detection limits and that is appropriate to the sample matrix or organism.

The variety and complexity of samples submitted for analysis using molecular biological techniques is vast. It is therefore beyond the scope of this chapter to offer a complete range of validated DNA extraction protocols. Equally, a small selection of protocols would have no more than a very limited relevance. Well developed protocols are perhaps most appropriately obtained from molecular biologists specialising in the sample of interest, often published in dedicated molecular biology manuals. Sources of a selection of DNA extraction techniques are given in Section 3.6. The information presented in this chapter aims to highlight generic validation issues that can arise at various stages in a broad range of DNA extraction procedures. In-house evaluation and optimisation of the relevant steps for each sample or matrix type can yield a reliable and valid technique.

Ideally, in an analytical environment, a good DNA extraction procedure should be as simple, safe and inexpensive as possible. It should also reproducibly furnish DNA of a sufficient quality and yield to allow subsequent analysis.

The suitability of isolated DNA as an analyte for a given technique is generally determined by three important factors: (i) amount or concentration,

(ii) purity and (iii) integrity of the DNA. Each of these factors can be influenced by the extraction technique employed and, in turn, impacts upon the validity of techniques applied in subsequent analysis.

- *Concentration or amount* — The amount of DNA obtained must be sufficient for all analyses, including relevant controls and duplications, to be carried out. The DNA must also be available at a workable concentration. Optimisation of the scale of the extraction technique, with respect to the required amount of analyte, will offer optimum reagent consumption and therefore better value for money.
- *Purity* — The isolated DNA should be as free as possible from all contaminants, both endogenous and exogenous, that may inhibit subsequent analyses. Where possible and appropriate, all contaminating nucleic acids should also be removed.
- *Integrity* — High molecular weight (HMW) DNA (ranging from 50 to 200 kbp) can be a requirement for some types of analyses, whereas in other cases a degree of DNA degradation can be tolerated. Should the DNA be sheared or degraded, the extent of such damage should be determined by agarose gel electrophoresis, in order to employ confidently the most applicable method of analysis.

3.2   The Various Steps of DNA Extraction

A generic DNA extraction process has to achieve a number of specific aims (which are set out below). In practice, an extraction technique may not necessarily consist of the sequential execution of each task. For example, extraction buffers commonly contain lysis chemicals as well as chemicals to chelate inhibitors. Depending on the sample type and subsequent analysis, greater or lesser emphasis may be placed on each of the following steps as required.

- *Sample preparation* — A sample may benefit from a number of preparative steps prior to cell or membrane lysis. This could include homogenisation, centrifugal separation or a step to minimise the effects of surface contaminants.
- *Cell or membrane lysis* — This step disrupts the cell wall/membrane and frees the DNA from cellular and organelle membranes. This can be accomplished by chemical (usually detergents), mechanical, enzymatic, microwave, sonication, heat or freeze/thaw treatments.
- *Protection and stabilisation of released DNA* — An extraction buffer is usually present during the lysis process. This contains a combination of chemical components which protect the released DNA in its new environment from degradation by cellular nucleases liberated during lysis.
- *Separation of nucleic acids from cell debris or sample matrix* — The separation of released DNA from cellular and matrix debris and other

biological macromolecules is traditionally achieved by phenol–chloroform and chloroform extractions.

- *Purification of DNA* — RNA may be removed from crude DNA extracts by the addition of appropriate nucleases. Other impurities that may act as inhibitors can, in some cases, be removed by appropriate ion exchange columns or the addition of chelating agents.
- *Concentration of DNA* — Both alcohol precipitation and commercial columns can concentrate the DNA to a suitable working molarity. Some inhibitors may also be removed at this stage.

3.3 Choosing the Most Appropriate DNA Extraction Procedure

Before a DNA extraction process is undertaken on a sample type for the first time, it is essential for the analyst to be informed of, and understand the consequences of, the following characteristics of the sample and the requirements of analysis.

1. The history of the sample:
- Is the sample unique? If so, DNA yields may need to be maximised to allow repeat or multiple analyses.
- Has the sample been maintained in a stable environment? This may help determine whether a method optimised for HMW or degraded DNA should be employed.
- Has the sample been exposed to a possible source of contamination? If so, certain sample preparation steps may be required that will remove contaminants.

2. The composition of the sample:
- Is the sample heterogeneous with respect to the analyte and matrix components? If so, preparation steps may be required to homogenise the sample or sampling issues should be addressed.
- Is the analyte contained in a matrix whose components are well characterised with respect to potential inhibitors or extraction strategies? If so, an existing extraction process could be adapted and optimised.
- Do the target organisms have cellular structures that lyse sufficiently under the conditions to be employed? This should be determined empirically.
- If multiple target organisms or tissues are being analysed simultaneously, will the lysis method used lyse all cells with equal efficiency? This is important if multiple target quantitation or detection is required from a single extraction, and should be determined empirically (e.g. Gram-positive and Gram-negative bacteria may lyse with different efficiency under the same conditions).
- Should the number of nuclei per mass of tissue or organism (i.e.

theoretical yield) be taken into account if quantitation is required? This
should be determined empirically.

3. *Time and resources available:*

- Should the sample be analysed by employing an extensive and possibly
  lengthy extraction process in order to maximise yield, purity and integrity,
  or, if time is of the essence, can these factors be selectively compromised
  and the sample adequately analysed by employing a shorter, somewhat
  cruder technique?
- Is a high throughput of samples required? If so, a technique with
  minimum hands-on time might be more appropriate if it does not
  compromise analytical quality.

4. *Subsequent analytical procedures:*

- Could endogenous chemicals (such as secondary metabolites) or
  exogenous compounds (such as food or fibre dyes) be present that inhibit
  or interfere with certain types of analyses? If so, additional purification
  steps may be required.
- Is quantitative analysis required? If so, appropriate steps to ensure sample
  homogeniety should be undertaken.
- Is HMW genomic DNA required or will degraded DNA be tolerated as
  an analyte? See Table 3.1.

Table 3.1 *DNA requirements of some common analytical techniques that employ genomic DNA*

Analytical method	DNA requirements		Approx. amount
	Purity	*Integrity*	
Gene specific PCR	High — crude tolerated	HMW or some degradation tolerated	1–5 ng/μl
RFLP fingerprinting	High	HMW required	20–100 ng
AFLP profiling	High	HMW but some degradation tolerated	10 ng/reaction
Arbitrarily primed PCR profiling (such as RAPD)	High	HMW required but some degradation tolerated	1–20 ng/μl
Dot or slot blot hybridisation	High	HMW but some degradation tolerated	100 ng–10 μg/dot
PFGE	High	very HMW required > 50 kbp	1–10 μg/sample
Cycle sequencing	High	Degradation tolerated	10 ng/μl
STR analysis	High	Degradation tolerated	0.1 ng/μl

An example of some of the points discussed in this section is given in Figure 3.1, where DNA was extracted from meat samples. Equal amounts of four types of bovine tissue (A, steak; B, kidney; C stewing steak; D, fat) were extracted using a typical validated method. At the stage prior to DNA precipitation, the extract was divided into four and precipitated as described before an aliquot was electrophoresed on an agarose gel and visualised under UV light. Figure 3.1 demonstrates how the alteration of a single procedure in the extraction process can alter the result obtained. In this instance, one of the four alcohol precipitations carried out (lanes 3) significantly reduced the yield of the DNA obtained. By comparing lanes A1, B1, C1 and D1 it can be seen that different DNA yields

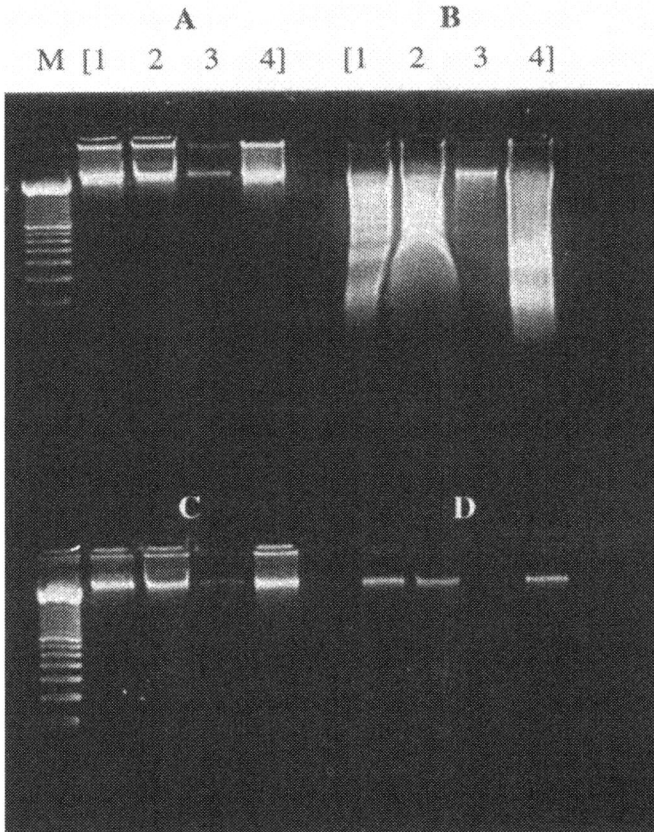

Figure 3.1   *Precipitation of DNA obtained from different mammalian tissue using standard ethanol and isopropanol procedures. Samples extracted using optimised DNA extraction method for meat. A, steak; B, kidney; C, stewing steak; D, fat. Lane 1, DNA precipitated with ethanol and ammonium acetate. Lane 2, DNA precipitated with ethanol. Lane 3, DNA precipitated with isopropanol and ammonium acetate. Lane 4, DNA precipitated with isopropanol. M, 100 bp molecular weight marker*

are obtained from the different tissue types, with reduced yields being associated with higher fat content samples. This could be due to differing cell lysis efficiencies or different cell sizes and therefore the number of nuclei present. It can also be seen that different qualities of DNA are obtained: the DNA obtained from the kidney tissue (B) shows signs of degradation, typically associated with programmed cell death; this could be due to poor storage conditions or tissue-specific enzyme activity.

An analyst prepared to ask the relevant questions, in possession of all the relevant information, and with relevant training, experience and understanding of the varied extraction processes, is in an excellent position to choose the most appropriate DNA extraction technique for a given sample.

In situations where analysis is confined to a single type of sample such as fresh blood, this decision may only have to be taken once, with reviews being made as the technology develops. However, some analytical laboratories are faced with the challenge of analysing DNA extracted from many different types of matrices. For example, the detection of adulteration in foodstuffs may involve identification of different meat species in various cooked products, the identification of different wheat species in dried pasta or the detection of animal matter in a dehydrated soup or sauce mix. In these cases the characteristics of the sample and analysis required, as discussed above, would have to be taken into account for each individual sample type before the most suitable extraction technique could be identified.

Under these circumstances it may be useful to take a unit-based approach to the DNA extraction process, breaking down the procedure into smaller, 'stand alone' units that could be validated in isolation. By applying a 'mix and match' selection of the most appropriate and efficient units of the extraction process, a tailor-made extraction process can be established. This approach can, when applied with due consideration and knowledge, reduce the need to revalidate an entire DNA extraction technique for each unique sample.

3.4   Validation Issues Arising at the Various Stages of DNA Extraction

The unit-based approach will also be adopted in this chapter for the evaluation of validation issues concerning DNA extraction techniques. Factors most likely to affect the yield, quality (purity and integrity), reproducibility and overall robustness of the various stages of an extraction process are discussed.

3.4.1   Sample Storage

Storage of the sample before submission for analysis can often be outside of the control of the analyst. Once in the laboratory, samples should be immediately placed in a stable and suitable environment. Such precautions will prevent contamination and further degradation of the sample.

Factors affecting the validity of DNA extraction due to sample storage have been identified as:

- *Incorrect sample storage temperature* — Further growth of contaminating microorganisms can be halted by freezing a sample. Naturally occurring autolysis and DNA degradation by endogenous and exogenous enzymes can also be abated by reduced temperatures. Samples requiring long-term storage should either be placed at $-80\,^{\circ}C$, preferably after snap cooling (fast freezing in liquid nitrogen), or lyophilised,[1] after which a sample can be stored at room temperature in a dry environment for extended periods. Any drying or freezing procedures should be undertaken speedily and as soon after the arrival of the sample as possible. Excessive freeze/thaw treatment of a sample should be avoided, as this may induce cell or DNA breakdown. Suitable precautions at this stage can help maintain the integrity of the DNA.
- *Incorrect sample storage environment* — Humidity can encourage microbial growth. Dust and aerosols originating from other samples can result in cross-contamination and the production of false positive results. Storage in a sterile, dry, sealed and sometimes dark environment may be necessary for samples placed at ambient or reduced temperatures.

3.4.2 Sample Preparation

A sample may require some manipulation in order to maximise the surface area to lysis forces ratio. In some circumstances it may be prudent to surface-clean or sterilise a sample, for example if the target organism is contained within a matrix or higher organism which could have been externally contaminated. Selective separation of the sample could also take place at this stage.

Factors affecting the validity of DNA extraction due to sample preparation have been identified as:

- *Homogeneity of sample* — It is essential that material selected for analysis is truly representative of the entire sample with respect to both matrix and analyte.[2] In order to achieve this, it may be necessary to homogenise heterogeneous samples in a blender or mortar and pestle before analysis. Alternatively, a minimum sample mass or size should be employed, thereby reducing the amount of misrepresentative sampling that may occur if a much smaller heterogeneous selection is made. Adherence to such empirically determined criteria can significantly improve reproducibility of yield, accuracy in quantitative analysis and avoid the possibility of false negative results.
- *Surface area to lysis forces ratio* — Bulky samples such as animal tissue may require breaking down into smaller segments or homogenisation in order to increase the efficiency of cell lysis. Failure to undertake this measure or to carry it out in an inconsistent manner may result in low or non-reproducible yields of DNA.
- *Cell or nucleic acid adherence to matrix material* — There is sometimes a requirement to remove cells from complex matrices or nuclei from cells before cell or membrane lysis takes place. In this way, inhibitors present

in the matrix or cell are removed at the beginning of the analysis. Caution should be exercised, however, as a percentage of cells and DNA may adhere to, or be entrapped within, certain matrices. This can certainly be the case with soil aggregates and some soil bacteria or free DNA. Consideration and avoidance of analyte loss can maximise yield, improve accuracy of quantitative analyses and maintain homogeneity of sampling.

- *Contamination* — Extreme care should be exercised to ensure that cross-contamination of samples does not occur during the early stages of analysis. Such occurrences could happen when a blender is used to homogenise multiple samples without adequate cleaning between specimens. Alternatively, the production of fine dust particles when grinding materials with liquid nitrogen should be suitably contained to avoid corruption of other samples. Protective clothing and gloves should be worn by all analysts to prevent bodily fluids or skin scales being inadvertently introduced into a sample and confusing the analysis.

3.4.3   Cell and Membrane Lysis

Cell lysis and the maintenance of DNA integrity are often conflicting aims of this part of the technique and sometimes a compromise has to be struck between maximising lysis activity and minimising shearing or degradation of the DNA. Several approaches can be taken to releasing nuclear DNA. The most common method is when both cell and membrane lysis is accomplished simultaneously under the same conditions.

When working with fresh plant and fungal material, it is sometimes preferable to achieve cell wall degradation first, followed by DNA isolation from protoplasts. In situations where inhibitory interference from cytoplasmic contaminants is to be avoided, intact nuclei can be isolated from cellular debris prior to lysis of the organelle or nuclear membrane. Although the last two approaches can be time consuming and are not generally suited to an analytical environment, they allow the genomic DNA to be released into an environment that minimises the degradation of the analyte and is relatively free of inhibitors.

A common cause of DNA degradation is endogenous DNases. The conditions for optimum activity of these nucleases vary from species to species; however, a number require magnesium as an essential co-factor (others may be dependent on other metallic co-factors even when present in the same organism,[3] and will only be active within a limited pH range). Addition of EDTA to the extraction buffer chelates magnesium ions, thereby reducing the chances of nuclease-induced DNA degradation. Optimisation of the extraction buffer pH and the addition of detergents may also minimise DNase activity. Oxidative damage of DNA can also occur particularly if samples are stored dried.[4]

Factors affecting the validity of cell or membrane lysis have been identified as:

- *Inaccessibility of cells to lysis forces* — If cells are not available to the enzyme or detergent, cell lysis will be inefficient, reducing yield and reproducibility (see Section 3.4.2).

Table 3.2 *Common detergents and denaturants used in DNA extraction*

Detergent/denaturant	*Description*	*Final concentration*
CTAB (cetyl-trimethyl-ammonium bromide)	A cationic detergent that solubilises membranes and forms a complex with DNA, allowing selective precipitation by lowering salt conc. <0.5 M, or adding isopropanol. May be useful for plants and matrices high in polysaccharides as these remain in solution (N.B. at higher salt concentrations contaminants are precipitated and DNA remains in solution)	1–2% (will precipitate out at <15 °C)
Guanidine isothiocyanate	A chaotropic agent and strong protein denaturant when used at high salt conc. Inhibits nucleases	4–5 M
Phenol	A protein denaturant. When added to crude aqueous DNA extracts, proteins collect in the organic phase or at the interface; DNA is maintained in the aqueous phase	1:1 to aqueous solution
Sarkosyl	An anionic detergent used instead of SDS due to its higher solubility (SDS is insoluble in high salt conc.)	0.5–2.0%
SDS (sodium dodecyl sulfate)	Anionic detergent and protein denaturant. Disassociates DNA–protein complexes. Foams easily when shaken	0.5–2.0%
Triton-X series	A series of gentle non-ionic detergents that solubilise proteins without denaturation	0.5%
Tween series	A series of gentle non-ionic detergents that solubilise proteins without denaturation	0.5%

- *Type and amount of detergent or denaturant used* — There are many different types of detergents and denaturants available for cell or membrane lysis (Table 3.2). Generally a detergent binds to the membrane and membrane lysis and solubilisation occurs to give detergent–lipid–protein complexes. Several factors can affect the performance of a given detergent including temperature, pH, ionic strength, detergent concentration, presence of multivalent ions and the presence of organic additives.[5] Therefore, if any of these factors are altered in a standard protocol, the effect on detergent activity should be empirically determined.
- *Concentration and activity of lytic enzyme* — Enzymes also perform at an optimum pH and temperature, and may be affected by metal ions or chelating agents. The concentration of enzyme used should therefore be empirically determined. Common enzymes used in cell and membrane lysis are listed in Table 3.3.
- *Concentration of EDTA in extraction buffer* — The concentration of

Table 3.3 *Common enzymes used in cell and membrane lysis*

Enzyme	Description	Approx. conc.
Proteinase K	A non-specific protease, not inactivated by metal ions or chelating agents. Full activity over pH 6.5–9.5. Frees nucleic acids from adhering proteins. Activity stimulated by denaturing agents (SDS and urea)	0.1–0.2 mg/ml
Pronase	A mixture of proteases; can be a cheaper alternative to proteinase K	0.5–1 mg/ml
Lysozyme	Used with EDTA to break down cell wall or membranes in bacterial DNA extractions	1–5 mg/ml
Zymolase/ chitinase	Digests chitinous cell walls of fungi that may be resistant to mechanical forces	1 mg/ml
Lyticase	Used for yeast cell wall degradation	20 units/ml

EDTA in the extraction buffer should be optimised to minimise the activity of endogenous DNases, thereby maintaining the integrity of the DNA. EDTA acts by chelating the co-factor magnesium and other divalent cations. For calcium-rich samples a specific complexing agent for calcium, EGTA, can also be added to the extraction buffer.

- *Concentration of salt in extraction buffer* — Salt creates an isotonic environment to stabilise free nucleic acids (phosphate buffered saline; 50 mM phosphate buffer, pH 7.4, 0.9% NaCl).
- *Extraction buffer pH* — The most common buffer used in molecular biology is Tris, with a pH range of 7.0–9.0. Other biological buffers are available which may be more suitable. For example, when working with acidic samples (e.g. certain foodstuffs) the pH may drop during the extraction procedure; therefore a buffer with a lower pH buffering range may be required. Extraction buffers should be carefully adjusted to the required pH. This should be carried out after all the components of the buffer are fully dissolved and, if required, the buffer has been autoclaved and cooled to room temperature. If stock extraction solutions are prepared (e.g. 10 × solution), the pH of the diluted, final working solution should be checked and adjusted as necessary.
- *Excessive damage of the DNA analyte* — Damage can be inflicted on DNA during the extraction process, causing shearing and degradation. DNA molecules are susceptible to fracturing if heated, sonicated, ground in liquid nitrogen or forced through small cavities such as pipette tips to an excessive degree. Where these procedures form an essential step in an extraction process, tolerance levels of such treatments should be determined empirically for a given sample. If high molecular weight DNA is required, all of the above should be avoided and gentler enzymatic lysis methods employed.

3.4.4 Separation of Nucleic Acids from Cell and Matrix Debris

Proteinaceous material is disrupted by proteinase K, denaturants and detergents in the extraction buffer. Deproteination and removal of debris is typically achieved by organic extraction with phenol–chloroform and chloroform, where the DNA remains in the aqueous phase and all debris and proteins either collect in the organic phase or sediment at the interface. In spite of the toxicity of phenol, organic separation is still extensively used. However, alternatives are now more frequently available, such as commercial DNA affinity columns which can eliminate extraction with organic solvents.

Factors affecting the validity of separation of nucleic acids have been identified as:

- *Phenol quality* — High quality phenol should always be used which has been correctly buffered (to pH 7–8) and stored in the dark at 4 °C for no more than 1 month or at −20 °C for longer periods. 8-Hydroxyquinoline (0.1%) stabilises phenol by retarding its oxidation. Phenol traces should be removed from the sample by a chloroform–isoamyl alcohol (24:1) extraction to avoid inhibition of subsequent enzyme activity.
- *Inefficient phenol extraction* — A sample may require more than one phenol extraction to remove potential contaminants. Care should be taken when removing the aqueous phase. The interface containing the debris should not be disturbed and it may be prudent to sacrifice a small volume of the aqueous phase close to the interface.

3.4.5 Additional Purification of DNA

Potential inhibitors of subsequent analyses can be removed at various stages of the extraction process. This is typically achieved in two ways. Firstly, reagents can be added to the extraction buffer that chelate or inactivate inhibitors. Secondly, an additional column-cleaning step can be included, which may either selectively bind DNA, allowing inhibitors to be eluted, or selectively bind inhibitors, allowing the DNA to be eluted.

Factors which can aid the additional purification of DNA have been identified as:

- *Composition of extraction buffer* — Certain reagents such as those given in Table 3.4 can be added to the extraction buffer in order to chelate or inactivate inhibitors.
- *Column cleaning* — Crude DNA extracts can be cleaned by passing the extract through a resin column. These can work in different ways, either by binding the DNA and washing through contaminating agents or by binding the potential contaminant and allowing the purified DNA to pass through. Common column matrices available for purifying DNA are listed in Table 3.5.

Table 3.4 *Reagent additives to extraction buffers that can remove potential inhibitors*

Additive	Reported action
PV(P)P 40 (polyvinyl(poly)pyrrolidine) (1–2%)	Included in buffers for extraction of plants and soils rich in polyphenols. Assists in the adsorption of phenolics. Polyphenols can be oxidised by phenol oxidases into compounds that form complexes with nucleic acids, causing damage to DNA and inhibit analyses involving enzymes. Should be prepared by acid wash[6]
DTT (dithiothreitol) (1 mM)	Antioxidant; considered superior to β-mercaptoethanol as it is odourless and has less of a tendency to be oxidised by air
DIECA (diethyldithiocarbamic acid) (4.0 mM/0.1%)	Phenol oxidase inhibitor. Inhibits the oxidation of polyphenols to quinonic substances that damage DNA
Ascorbic acid (5 mM)	Strong reducing agent/antioxidant
β-Mercaptoethanol (0.2–0.5%)	Antioxidant/reducing agent; protects thiol groups of enzymes against oxidation. Add to extraction buffer just before use
Cysteine (10 mM)	Antioxidant

Table 3.5 *Common column matrices available for purifying DNA (used according to the manufacturer's instructions)*

Column/matrix	Reported action
Hydroxyapatite	A form of calcium phosphate that binds dsDNA selectively in a mixture of nucleic acid types
Gel filtration	Purifies by size fractionation. Examples include Sephadex G50–G200 and CL6B separose, which excludes smaller fragments (<194 bp). Spin column formats are available
Silica particles	Absorbs DNA; examples include glassmilk/glass fog. Qiagen surface modified silica gel acts as an anion exchange resin
Ultrafiltration	Membrane capture of DNA while smaller contaminants pass through. Examples include Centricon® concentrator (Amicon)
Anion exchange resins based on cellulose, dextran or agarose (sold under various commercial names)	Binds the strongly anionic DNA. The use of anionic detergents such as SDS should be avoided as these will also bind. Examples include DEAE sephacel/separose/sephadex
Cation exchange resins (Chelex® 100, BioRad)	Chelating resin binds metals and other potential inhibitors; DNA should not be stored long term as inhibitors may be released over time. Yields partially single stranded DNA; therefore may give bias to quantification using intercalating dyes

- *RNase treatment of the sample* — RNA contamination can interfere with some analytical procedures or quantitation techniques. This may be particularly relevant when working with RNA-rich tissues such as liver and kidney. Digestion of RNA can be undertaken at various stages of the extraction process if the buffer conditions are suitable. Care must be taken to inactivate DNases present in commercial RNase preparations according to the manufacturer's instructions. A stock solution of 10 mg/ml is normally prepared and used at a working concentration of approximately 100 μg/ml.

3.4.6 Precipitation and Concentration of DNA

This step in the procedure can be carried out for a number of reasons. Firstly, to change the solvent, perhaps to a suitable buffer for storage purposes; secondly, to remove certain non-precipitated contaminants; and thirdly, to concentrate the DNA to a required working molarity. Optimisation of this stage in the extraction process is perhaps the most neglected step. Additional attention at this point could enhance both reproducibility and yield of extracted DNA. A review of the practical aspects of alcohol precipitation is given by Winfrey *et al.*[7]

Factors affecting the validity of DNA extraction due to the concentration or precipitation of DNA have been identified as:

- *Volume and temperature of alcohol used and precipitation times* — Either a 2× volume of ethanol or a 0.6× volume of isopropanol is usually recommended. DNA precipitations are undertaken at room temperature or at $-20\,^\circ$C. Yields can be increased by overnight incubations at $-20\,^\circ$C, while for samples known to contain high quantities of DNA, 10 min at room temperature may be sufficient. Centrifugation times and speed can also be increased for low yield samples; however, this may make the pellet more compact and harder to redissolve. Isopropanol is useful when smaller volumes are required; however, it is less volatile and has more of a tendency to co-precipitate salts than ethanol.
- *Concentration and type of salt* — In order for a nucleic acid precipitate to form, there must be at least 0.2 M concentration of a monovalent cation to shield the negative charge of the nucleic acid phosphate groups and allow aggregation of nucleic acid strands.[7] Therefore, the addition of salt to samples extracted with common buffers, which have comparatively low salt concentrations, may help the alcohol precipitation process if added to the final concentrations given in Table 3.6. However, if a high salt concentration buffer were used, the addition of further salts could reduce the yield of precipitated DNA. Washing the pellet with 70% ethanol can remove some of the precipitated salts.
- *Degraded DNA* — When degraded DNA is expected from an extraction process (e.g. when working with a highly processed sample), it should be remembered that short molecules of DNA (<200 bp) are precipitated

Table 3.6 *Recommended salt concentrations used in alcohol precipitations*

Salt	Stock concentration	Final concentration	Specific use
Sodium acetate	3.0 M, pH 5.2	0.3 M	Generally used for DNA
Ammonium acetate	7.5 M	2.5 M	Reduces co-precipitation of free dNTPs
Sodium chloride	5 M	0.1 M	Used for samples containing SDS (SDS remains in solution)

inefficiently by ethanol. Precipitation can be improved by the addition of $MgCl_2$ or glycogen to a final concentration of 10 mM and 10 $\mu g/ml$, respectively.

- *DNA amount too low* — If small-scale or low-yield extraction procedures are undertaken, DNA could be lost at the precipitation step. Addition of glycogen (to a final concentration of 10 $\mu g/ml$) can act as an efficient, inert co-precipitant of low concentrations of DNA. Commercial columns for concentrating DNA, such as Centricon® ultrafiltration tubes, could be used as an alternative to alcohol precipitation. Correct choice of unit with the appropriate nucleotide cut-off is required to avoid loss of degraded DNA.
- *Pellet loss* — Centrifugation tubes should be placed consistently in a centrifuge in such a way that the position of the DNA pellet is always known. In this way, even when the pellet is translucent and difficult to see, contact with the pellet, possibly causing it to dislodge, can be avoided.
- *Pellet not resuspended completely* — High molecular weight DNA might take several hours to redissolve completely and pure DNA may dissolve more quickly than contaminated DNA pellets. Vigorous mixing or vortexing at this stage could damage the integrity of the DNA. In order to obtain maximum yield, integrity and reproducibility of DNA in the extraction process, pellets should be resuspended at 4 °C overnight without agitation. If this still does not resuspend the pellet, resuspension can be undertaken by placing the sample at 45–50 °C for 15–20 min with occasional gentle shaking. It should also be remembered that up to 50% of the DNA may be smeared up the edge of the tube rather than being contained in the pellet.
- *Alcohol precipitated inhibitors* — If it is found that inhibitors of certain analytical techniques are also precipitated, alternative methods of DNA concentration should be investigated such as ultrafiltration columns (Table 3.5). For high yield extractions, DNA can be removed by spooling or collecting the DNA precipitate on a sterile glass hook. Some of the salts or inhibitors that are precipitated by 95% ethanol may be removed by a 70% ethanol wash.

Table 3.7 *DNA extraction methodologies*

Special application	Type of extraction and source
Range of plants Fresh or hydrated plant tissue	CTAB[9]
Removes contaminated proteins and polysaccharides from plant material	Hot SDS[9]
Fresh plant material only	Isolation of nuclei followed by CTAB[9]
Suitable for 'difficult' plants high in polysaccharides and polyphenols Low yields	High potassium acetate[9]
Various plant tissues, herbarium samples and woody plants Rapid extraction process	CTAB/PVP/ascorbic acid/DIECA followed by chloroform/IAA[10]
Plants high in polysaccharides	CTAB/PVP/chloroform[11]
Small scale and rapid methods for plants	Comparison of methods[12]
Plants with high phenolic and tannic content	Citrate buffer/Triton X glucose-nuclei isolation[13]
Various, including difficult plant species	Various[14]
From mammalian tissue sample	SDS/proteinase K/phenol/chloroform[15]
From uncoagulated blood	NH_4Cl/SDS/proteinase K/salting out[15]
From mammalian cell cultures	Nonidet/DTT followed by phenol or salting out extraction[15]
Blood/tissue samples	SDS/proteinase K/phenol/chloroform[15]
Mammalian tissues rich in mucopolysaccharide	CTAB/PVP/chloroform[15]
Ancient bone tissue	Comparison of three different techniques[16]
Biological evidence material (blood stains, hair, tissue, sperm cells and epithelial cells)	Various techniques[17]
Blood, body fluids and stains (FBI protocols)	SSC/SDS/proteinase K/phenol[18] Saline/sarcosyl/proteinase K[18]
Cigarette butts, postage stamps and other saliva stained material	SDS/proteinase K[19]
Blood and forensic samples	Chelex®[20]
Fungal mycelium or spores	CTAB followed by chloroform/IAA[21]
Histoplasma capsulatum	SDS/boiling followed by phenol/chloroform[22]
Fresh and herbarium fungal samples	Crushing/chelating/detergent methods[23]
Various fungi	Various[14]
Soil microbes	Various[24]
Food-borne microbial pathogens from cheeses, milk and meats	Various[25]
Microbes from soil, sediment and aquatic environments	Various[26]
Microbes from clinical samples including gastric biopsy, clinical swabs, saliva, stool and blood samples	Various[27]
Bird feathers	Chelex-100/proteinase K[28]
Insects	CTAB/proteinase K[29]
Nucleic acids from aquatic environments	Various[30]

3.5 Summary

Traditional DNA extraction methodologies employing chemicals such as SDS, proteinase K and phenol are now reasonably well established.[8] These tend, however, to be time consuming and involve multiple liquid transfer operations. Alternative commercial kits often offer reduced hands-on time and cleaner approaches to extractions; however, these can be more expensive and limited to very specific applications. In an analytical environment, neither of the approaches mentioned may be ideal and the situation can be further complicated by the sample matrix composition. Where simple approaches may be preferable, complex matrices and non-ideal samples may demand additional clean-up procedures. Measurement of DNA yield in itself is not sufficient to determine the suitability of an extraction methodology. The quality, encompassing purity and integrity of the DNA analyte can also be of paramount importance. An inappropriate choice or a sub-optimal extraction methodology could have significant consequences on subsequent analyses, which may have to be repeated or produce false negative results. Validation of sampling procedures, sample storage, sample preparation and DNA extraction should all be considered vital to the production of quality data in subsequent analyses.

3.6   Genomic DNA Extraction Protocols

This section provides the reader with some sources of well developed DNA extraction protocols (Table 3.7). Generally, the references given are excellent review articles and most contain many other sources of genomic DNA extraction procedures. These articles were chosen as they often offer an overview of the challenges associated with the analysis of specialised materials, are written by experts in a given field, and in many cases offer valuable troubleshooting advice.

3.7   References

1. Day, J. G. and McLellan, M. R. (eds.) 1995. Cryopreservation and Freeze-drying Protocols. Humana Press, Totowa, NJ.
2. Crosby, N. T. and Patel, I. 1995. General Principles of Good Sampling Practice. The Royal Society of Chemistry. Cambridge.
3. Kokilera, L. 1995. Comparative study of induction of endogenous DNA degradation in rat liver nuclei and intact thymocytes. *Comp. Biochem. Physiol.* **111B**: 35–43.
4. Matsuo, S., Toyokuni, S., Osaka, M., Hamazaki, S. and Sugiyama, T. 1995. Degradation of DNA in dried tissues by atmospheric oxygen. *Biochem. Biophys. Res. Commun.* **208**: 1021–1027.
5. Neugebauer, J. 1994. A Guide to the Properties and Uses of Detergents in Biology and Biochemistry. Calbiochem-Novabiochem International, Trade Publication.
6. Rochelle, P. A., Will, J. A. K., Fry, J. C., Jenkins, G. J. S., Parkes, R. J., Turley, C. M. and Weightman, A. J. 1995. Extraction and amplification of 16S rRNA genes from deep marine sediments and seawater to assess bacterial community diversity. In: Nucleic Acids in the Environment (eds. Trevors, J. T. and van Elsas, J. D.), pp. 219–240. Springer, Berlin.
7. Winfrey, M. R., Rott, M. A. and Wortman, A. T. 1997. Appendix VII. Alcohol

precipitation of nucleic acids. In: Unraveling DNA. Molecular Biology for the Laboratory. Prentice-Hall, Englewood Cliffs, NJ.

8. Sambrook, J., Fritsch, E. F. and Maniatis, T. 1989. Molecular Cloning. A Laboratory Manual, Cold Spring Harbor, New York.
9. Wilke, S. 1997. Isolation of total genomic DNA. In: Plant Molecular Biology — A Laboratory Manual (ed. Clark, M. S.), pp. 3–14. Springer, Berlin.
10. Neal Stewart, C., Jr. 1997. Rapid DNA extraction from plants. In: Fingerprinting Methods Based on Arbitrarily Primed PCR (eds. Micheli, M. R. and Bova, R.), pp. 25–28. Springer, Berlin.
11. Towner, P. 1991. Purification of DNA. In: Essential Molecular Biology, vol. I (ed. Brown, T. A.), pp. 47–55. Oxford University Press, New York.
12. Rogers, H. J., Burns, N. A. and Parkes, H. C. 1996. Comparison of small scale methods for the rapid extraction of plant DNA suitable for PCR analysis. *Plant Mol. Biol. Reporter* **14**: 170–183.
13. Katterman, F. R. H. and Shattuck, V. I. 1983. An effective method of DNA isolation from the mature leaves of *Gossypium* species that contain large amounts of phenolic terpenoids and tannins. *Prep. Biochem.* **13**: 347–359.
14. Weising, K., Nybom, H., Wolff, K. and Meyer, W. 1995. Methodology. In: DNA Fingerprinting in Plants and Fungi, pp. 43–74. CRC Press, Boca Raton, FL.
15. D'Ambrosio, E. and Pascale, E. 1997. DNA extraction from mammals. In: Fingerprinting Methods Based on Arbitrarily Primed PCR (eds. Micheli, M. R. and Bova, R.), pp. 15–20. Springer, Berlin.
16. Nielsen, H., Engberg, J. and Thuesen, I. 1994. DNA from arctic human burials. In: Ancient DNA (eds. Herrman, B. and Hummel, S.), pp. 122–140. Springer, New York.
17. Sensabaugh, G. F. 1994. DNA typing of biological evidence material. In: Ancient DNA (eds. Herrman, B. and Hummel, S.), pp. 141–148. Springer, New York.
18. Easteal, S., McLeod, N. and Reed, K. 1991. Appendix C. In: DNA Profiling. Principles, Pitfalls and Potential. Harwood Academic Publishers, Switzerland.
19. Hochmeister, M. N., Rudin, O. and Anbach, E. 1998. PCR analysis from cigarette butts, postage stamps, envelope sealing flaps and other saliva-stained material. In: Forensic DNA Profiling Protocols (eds. Lincoln, P. J. and Thomson, J.), pp. 27–32. Humana Press, Totowa, NJ.
20. Willard, J. M., Lee, D. A. and Holland, M. M. 1998. Recovery of DNA for PCR amplification from blood and forensic samples using a chelating resin. In: Forensic DNA Profiling Protocols (eds. Lincoln, P. J. and Thomson, J.), pp. 9–18. Humana Press, Totowa, NJ.
21. Graham, G. C. and Henry, R. J. 1997. Preparation of fungal genomic DNA for PCR. In: Fingerprinting Methods Based on Arbitrarily Primed PCR (eds. Micheli, M. R. and Bova, R.), pp. 21–24. Springer, Berlin.
22. Woods, J. P., Berg, D. E. and Fisher, K. L. 1997. Extraction of *Histoplasma capsulatum* DNA for PCR. In: Fingerprinting Methods Based on Arbitrarily Primed PCR (eds. Micheli, M. R. and Bova, R.), pp. 35–40. Springer, Berlin.
23. Taylor, J. W. and Swann, E. C. 1994. DNA from herbarium specimens. In: Ancient DNA (eds. Herrman, B. and Hummel, S.), pp. 166–181. Springer, New York.
24. Saano, A., Tas, E., Pippola, S., Lindstrom, K. and van Elsas, J. D. 1995. Extraction and analysis of microbial DNA from soil. In: Nucleic Acids in the Environment (eds. Trevors, J. T. and van Elsas, J. D.), pp. 49–68. Springer, Berlin.
25. Jones, D. D. and Bej, A. K. 1994. Detection of foodborne microbial pathogens using polymerase chain reaction. In: PCR Technology: Current Innovations (eds. Griffin, H. G. and Griffin, A. M.), pp. 341–362. CRC Press, Boca Raton, FL.
26. Bej, A. K. and Mahbubani, M. H. 1994. Applications of the polymerase chain reaction (PCR) in vitro DNA-amplification method in environmental microbiology. In: PCR Technology: Current Innovations (eds. Griffin, H. G. and Griffin, A. M.), pp. 327–339. CRC Press, Boca Raton, FL.

27. Mahbubani, M. H. and Bej, A. K. 1994. Application of polymerase chain reaction methodology in clinical diagnostics. In: PCR Technology: Current Innovations (eds. Griffin, H. G. and Griffin, A. M.), pp. 307–326. CRC Press, Boca Raton, FL.
28. Ellegren, H. 1994. Genomic DNA from museum bird feathers. In: Ancient DNA (eds. Herrman, B. and Hummel, S.), pp. 211–217. Springer, New York.
29. Hunt, G. J. 1996. Insect DNA extraction protocol. In: Fingerprinting Methods Based on Arbitrarily Primed PCR (eds. Micheli, M. R. and Bova, R.), pp. 29–34. Springer, Berlin.
30. Paul, J. H. and Pichard, S. L. 1995. Extraction of DNA and RNA from aquatic environments. In: Nucleic Acids in the Environment (eds. Trevors, J. T. and van Elsas, J. D.), pp. 153–178. Springer, Berlin.

CHAPTER 4

Quantification of Total DNA by Spectroscopy

PAUL A. HEATON

4.1 Introduction

Quantification of DNA is becoming increasingly important in analytical molecular biology as the technology strives to offer increasingly sensitive and informative data. Quantification of total DNA is required for several reasons:

- To determine the yield obtained from a DNA extraction process for comparative and validation purposes.
- To determine the amount of starting template (target molecule quantification) for analytical procedures such as PCR (including its many variations), hybridisation and cycle sequencing. The routine use of such techniques depends heavily on the accurate and reproducible determination of the amount of starting DNA in order to enhance the reproducibility and quality of the data obtained.
- To determine the amount of a target genome or relative amounts of target in a mixed sample, e.g. detection of adulterated foodstuffs[1] and detection of human DNA in microbially contaminated samples.[2] Quantification of DNA for these purposes can be achieved by employing slot or dot blot hybridisation[3] (Chapter 9).

Several methods exist which can be used to quantify DNA.[4,5] Perhaps one of the most commonly used and simplest techniques is spectrophotometric determinations, either UV absorption or fluorescent emission. This chapter will attempt to bring to the analyst's attention the potential sources of error in determining DNA concentration using this approach. The general procedure to determine a DNA concentration by spectroscopy can be divided into two steps: firstly, the absorbance or fluorescence of the sample is measured; and secondly, this spectroscopically determined value is converted into a DNA concentration. Though this procedure appears simple, practically, it can be fraught with problems. For example, the presence of DNA degradation products or contaminants in the DNA sample, which absorb in the same range as DNA or quench fluorescence, will affect the spectroscopic measurements made. In addition, the choice of conversion factors used to transform these values into DNA concentration measurements can influence the result obtained. This chapter describes the effects of various contaminants on DNA concentration determinations, methods to remove these contaminants, the effect of DNA degradation and methods to convert spectroscopic measurements into DNA concentration values.

4.2   Determining DNA Concentration by Ultraviolet Spectroscopy

All nucleic acids absorb strongly in the UV region with a maximum occurring at a wavelength of 260 nm and this physical property can be used as a basis for a method to determine the concentration of nucleic acids in solution. Measurement of DNA concentrations by UV absorbance is a relatively simple procedure. This involves:

1. Diluting a portion of a DNA-containing solution, usually in double distilled H_2O (ddH_2O) or TE buffer, so that its absorbance at 260 nm is less than 1 optical density (OD) unit.
2. Determining the absorbance value of the sample, which is then adjusted for the dilution factor.
3. Conversion of the adjusted absorbance into a concentration, in $\mu g/ml$ or μM, using a simple conversion factor, which can either be determined from a calibration graph or approximated.

Accurate UV absorbance measurements can be made down to at least 0.01 OD units, which corresponds to a lower DNA concentration limit of approximately 0.5 $\mu g/ml$. The main drawback with determining DNA concentrations by UV is specificity: any contaminant absorbing at 260 nm will contribute to the final DNA concentration, leading to an overestimation. Potential common contaminants which absorb in this range are RNA, proteins and phenol.

RNA contamination is a particular problem since its absorbance spectrum is practically indistinguishable from that of DNA, making potential contaminations difficult to detect. Thus any contaminating RNA with a concentration in excess of 0.4 $\mu g/ml$, corresponding to an A_{260} of 0.01 OD, will affect the final DNA concentration. RNA contaminations can be removed simply by digestion with RNase A followed by a purification step to remove the protein (further details may be found in Chapter 3).

Proteins, in contrast to RNA, have a UV absorbance spectrum that is easily distinguished from DNA. Typically, proteins have a maximum at 280 nm with some absorbance at 260 nm. A protein concentration of 0.3 mg/ml has an absorbance of around 0.01 OD at 260 nm and this value can be taken as an upper limit before an effect on the determined DNA concentration is observed.

A simple but rather insensitive method to detect a protein contamination in a sample is to determine the $A_{260}:A_{280}$ ratio. In an uncontaminated sample of DNA, this ratio should be $1.8:2.0$;[6] contamination with protein results in lower values; however, any deviation from this could indicate a contamination problem. Care must be taken when using the $A_{260}:A_{280}$ ratio as an assessment of sample purity as the ratio is also dependent on the pH and ionic strength of the sample.[7] As pH increases, the absorbance at 280 nm decreases but the absorbance at 260 nm is unaffected, which results in the ratio having elevated values at higher pH. In contrast, increasing the ionic strength, i.e. the salt concentration, tends to decrease both absorbancies but has the overall effect of increasing the $A_{260}:A_{280}$ ratio. Protein contamination can be removed by digestion with proteinase K followed by a purification step to remove the protein digestion products (further details may be found in Chapter 3).

Phenol, like proteins, has an absorbance spectrum which is readily distinguishable from nucleic acids: it has an absorbance maximum at 264 nm and thus also absorbs strongly at 260 and 280 nm. An approximate $1:100\,000$ dilution of tris-saturated phenol has an A_{260} of 0.01 OD and this can be considered a lower detection limit. As with proteins, the $A_{260}:A_{280}$ ratio can be used to indicate a phenol contamination problem. Any residual phenol may be removed by a simple chloroform extraction.

The conversion of A_{260} values to concentrations are based on the Beer–Lambert Law (eqn. 1). This states that the measured absorbance (A) of a DNA solution is determined by the DNA concentration (c), the path length (l) and the extinction coefficient (ε). This law, which predicts a linear change in absorbance with concentration, is widely applicable, but this linear relationship does break down at high DNA concentrations. The path length is usually fixed at 1 cm by the spectrophotometer, whereas the value of the extinction coefficient (ε) may be either calculated, determined experimentally or, more usually, approximated.

$$A = \varepsilon c l \tag{1}$$

Determination of sample DNA concentration using UV spectrophotometry can also be estimated by comparison to DNA quantification standards through the preparation of a calibration graph (Section 4.3.2).

4.2.1 Determining the Extinction Coefficient

Calculation of ε — The value of the extinction coefficient for a particular DNA is dependent on the wavelength at which it is measured, which is typically 260 nm, and the length and sequence of the nucleic acid. Each component base in the DNA contributes to ε, but the contribution is not simply additive since neighbouring bases interact with each other, reducing their extinction coefficients at 260 nm (the hypochromic shift). Calculation of ε can be adopted for short oligonucleotides; however, different approaches are usually adopted for genomic DNA due to its complexity.

Experimental determination of ε — The most common method is to compare the absorbance of a given DNA sample before (A_0) and after (A_∞) total enzymatic digestion, for example with snake venom phosphodiesterase, to its component 5'-mononucleotides.[8] The ratio of $A_0:A_\infty$ can then be used in conjunction with the sum of the extinction coefficients of each of the separate bases (ε_{sum}) to calculate ε for the intact DNA under test (eqn. 2). Again, the size and sequence of the DNA under test is required.

$$\varepsilon_{intact} = \frac{A_0}{A_\infty}\varepsilon_{sum} \qquad (2)$$

Approximation of ε — Usually, the length and sequence of an extracted genomic DNA is not available, which makes the accurate determination of the extinction coefficient impossible. It is at this point that problems arise. In these situations, suitable approximations need to be made. Typical values for ε *per base* are 6600 M^{-1} cm^{-1} for dsDNA and dsRNA and 8500 M^{-1} cm^{-1} for single-stranded nucleic acids.[5] From these values of ε, the much quoted values '50 μg/ml dsDNA or 40 μg/ml ssDNA have an absorbance reading of 1' are derived.[6]

Whichever method is used to determine ε, eqn. (3) can be used to calculate the DNA concentration in a sample:

$$c = \frac{\text{absorbance of DNA sample}}{\text{extinction coefficient of the DNA}}$$

i.e.

$$c = \frac{A \cdot d}{\varepsilon_{per\ base} \cdot n} \qquad (3)$$

where A is the measured absorbance, d is the dilution factor and n the number of bases in the DNA.

4.2.2 Practical Aspects of Measuring DNA Concentrations by UV Spectroscopy

Other factors, apart from ε, have an influence on the validity of the DNA concentration determined using UV spectrophotometry. These include the following:

- *Calibration of the spectrophotometer* — Calibration should be performed regularly using standard filters which are transparent to only certain defined wavelengths. This can be performed either in-house or by external service personnel. Low $A_{260}:A_{280}$ ratio values can, in some cases, indicate a problem with the spectrophotometer.[9]
- *Cuvettes* — These should be of good quality and transparent in the UV region. Care should be taken to handle cuvettes only on their non-optical surfaces. Cuvettes should be rinsed with 95% ethanol followed by ddH$_2$O and wiped dry with lint-free paper tissue. Mis-matching of cuvettes can

sometimes occur; to test for this, cuvettes should be filled with blank solutions and absorbancies read and compared. If a significant and consistent difference is obtained, appropriate adjustment of sample readings can be made.

- *Sample preparation* — Care should be taken to ensure that DNA in the samples under test is fully dissolved before concentrations are measured. In addition, particulate matter, if present, can be sedimented by centrifuging the DNA-containing sample and decanting the solution, or filtering the sample through a piece of tissue paper in a Pasteur pipette.
- *Reference blank* — This is a cuvette containing an identical solution to the cuvette which is being assayed, except the DNA component is excluded. A blank should always be used when measuring DNA concentrations to correct for any absorbance of the solution containing the DNA.
- *Sample dilution* — Absorbance (A), and therefore concentration, are related to the logarithm of transmittance (T) of the sample, which is defined as the ratio of light passing through the sample (P_{sample}) compared with the blank (P_{blank}) (eqn. 4). As a result of this relationship, errors in the measured transmittance at low concentrations have a far larger effect on the concentration measurement than those measured at higher DNA concentrations. Thus, when determining DNA concentrations using the spectrophotometer, a higher DNA concentration is preferable so long as the absorbance readings are in the linear range, which is usually at absorbance values of < 1.

$$A = -\log T \quad \text{where} \quad T = P_{sample}/P_{blank} \quad (4)$$

- *Light source* — The deuterium lamp in the spectrophotometer should be allowed to warm-up before use (according to the manufacturer's instructions) to allow the emissions from the source to stabilise.

4.3 Determination of DNA Concentration by Fluorescence Spectroscopy

Determination of DNA concentration by fluorescence spectroscopy relies on the fluorescent enhancement of a dye on intercalation into a nucleic acid. An increasing range of dyes is becoming available with improved sensitivities, some of which bind specifically to a particular type of nucleic acid, such ds- and ssDNA and RNA. Measurements made with such dyes are thus free from interference from other nucleic acid contaminants, which can give erroneous measurements when using UV spectroscopy. Concentration measurements are made in a similar manner to the UV method:

1. The DNA is diluted to a suitable concentration in ddH$_2$O or a buffer, such as TE, so that, on addition of the fluorescent dye, its fluorescence is within the measurement range of the fluorimeter.

2. A fluorescent dye is added, which should be at a concentration sufficient to saturate all possible intercalation sites on the DNA analyte and should be kept at a fixed concentration for the series of measurements.
3. A fluorescence measurement is made.
4. The fluorescence value is then be converted into a DNA concentration using an empirically determined conversion factor.

4.3.1 Practical Aspects of Measuring DNA Concentrations by Fluorescence Spectroscopy

Other factors which should be considered when determining DNA concentrations by fluorescence include:

- *Sample preparation* — The DNA sample should be fully dissolved and free from particulate material before measurements are made.
- *Reference blank* — Reference blanks should be used, which have the same composition as the test samples except with the exclusion of the DNA component.
- *Selecting the dye* — A dye should be selected which is sensitive enough to detect DNA concentrations in the range anticipated in the test samples and if possible be selective for the nucleic acid of interest.
- *Dye concentration* — The amount of dye used should be constant in the test samples and the DNA standard used to construct the calibration curve. This amount should be sufficient to saturate the range of DNA under test. The dye saturation point may be determined empirically by measuring the fluorescence of samples containing a fixed amount of dye and a varying amount of DNA. The point at which the plot deviates from linearity is the point at which DNA saturation has been exceeded.
- *Microtitre plates* — When using a microtitre format and plate reader to measure fluorescence, black-walled plates are recommended to minimise interference from adjacent wells when measurements are being made.

4.3.2 Preparation of a Calibration Graph

The fluorescent conversion factor is determined by means of a calibration graph, on which the fluorescence of a serial dilution of DNA standards of known concentration has been plotted against quantity.[10] Known concentrations could have been measured by some other established methodology, i.e. UV spectroscopy, or supplied by the manufacturer. Assessing the exact amount of nucleic acid for standards is, however, a major challenge for absolute quantification purposes; therefore consistency in the source and use of standards is of paramount importance. As expected, the nature of the DNA standard used for the calibration graph and the conditions under which its fluorescence is measured are critical to the determination of accurate and reliable DNA concentrations.

- *The DNA standard* — The fluorescent intensity of a given DNA, after intercalation of a fluorescent dye, is dependent on the length and, in some instances the sequence, of the DNA used. Ideally, the DNA standard used for the serial dilution should be pure target DNA, but this is not always feasible. In such circumstances, a compromise needs to be made and a commercially available pure DNA (e.g. ultrapure calf thymus DNA) of similar length may be used. In this case, it is assumed that a similar number of dye molecules intercalate into the selected standard DNA as into the target, i.e. the dye does not display any sequence specificity.
- *Measurement conditions* — The conditions under which the standard DNA is prepared and measured must be identical to those used for the sample DNA. Important factors include final concentration of dye, the buffer used to dilute the samples, final salt concentrations, and the fluorometer settings. The range of standard concentrations used should span the expected range of sample concentrations, with the majority of the samples falling in the mid-range. Care must be taken when fluorescence measurements are made in microtitre plates. As the wells on the plate are quite close together, fluorescence measurements can be affected by the fluorescence induced in neighbouring wells. To avoid this, opaque walled microtitre plates are used.

Once a suitable DNA standard has been selected and a series of concentrations measured, a standard calibration graph may be plotted. The variation in fluorescence with concentration should be linear provided the DNA concentration is not excessively high and sufficient dye has been used to saturate all possible intercalation sites. The gradient of the calibration graph can then be used as a conversion factor to determine sample DNA concentrations.

4.3.3 Fluorescent Dyes

Many fluorescent dyes are commercially available and a good source is Molecular Probes, Eugene, Oregon. Three dyes are highlighted below: ethidium bromide, PicoGreen and Hoechst 33258.

4.3.3.1 *Ethidium Bromide*

Although ethidium bromide is not usually employed with spectrophotometric techniques, it is the traditional reagent for the fluorescent labelling of nucleic acids and is therefore included as one of the major alternatives to spectrophotometry. It has a broad specificity in that it will label both DNA and RNA, but it displays a distinct preference for dsDNA. Ethidium bromide has a limited sensitivity, down to ≈ 100 ng/ml, since it has only a moderate affinity for nucleic acids and moderate fluorescent enhancement on binding. Using ethidium bromide, samples can be stained after or during agarose gel electrophoresis. The fluorescence is induced by UV light and is proportional to the amount of DNA. Comparison to a set of standards of known concentrations (usually lambda DNA or molecular weight markers) gives an estimation of DNA quantity in ng/μl. Although perhaps more time consuming than spectroscopy measurements, this approach is included in the text as it is commonly used and has some distinct advantages. It allows the simultaneous semi-quantification of genomic DNA, estimation of RNA contamination and an evaluation of DNA integrity. Ethidium bromide is toxic and is suspected to cause inheritable genetic damage; thus suitable precautions must be taken when handling this material. Disposal methods for solutions containing ethidium bromide are available.[11]

4.3.3.2 PicoGreen

PicoGreen is a fluorescent dye which displays excitation and emission wavelengths in a similar range to fluorescein (480 nm and 520 nm, respectively). It also exhibits very specific binding to dsDNA but is unaffected by ssDNA, RNA and protein contaminants.[12] Since PicoGreen has a very high affinity for dsDNA and shows a high fluorescent enhancement on binding, the dye can be used to detect dsDNA in the range 25 pg/ml to 1 μg/ml.[12]

When using PicoGreen, care must be taken to control the conditions under which fluorescence measurements are made (Table 4.1). It is important to ensure that divalent metal ions, such as Mg^{2+}, Ca^{2+} and Zn^{2+}, are not present, as these ions have been shown to have a strong quenching effect on PicoGreen fluorescence[12] and the addition of EDTA to chelate these ions has a minimal effect in alleviating this (Figures 4.1 and 4.2). In contrast, monovalent ions such

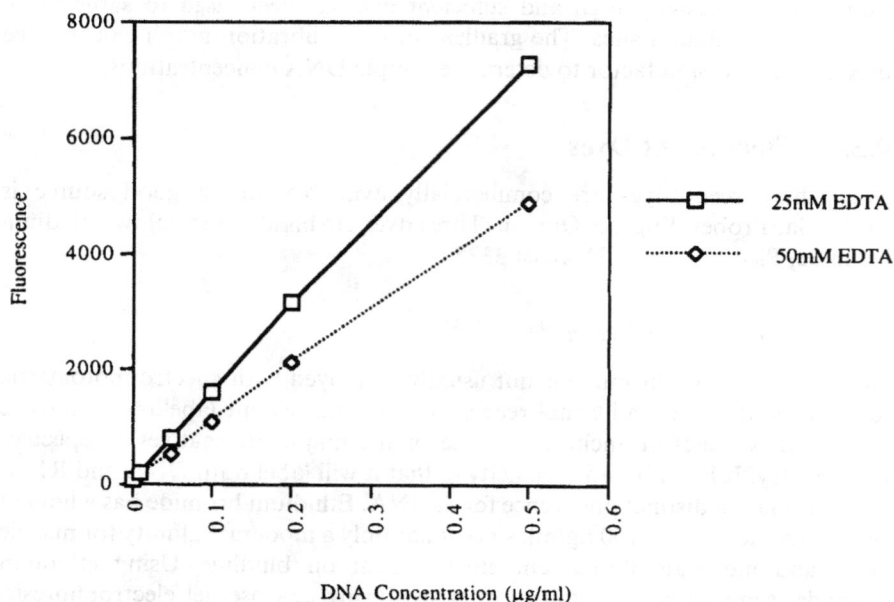

Figure 4.1   *Effect of EDTA on the fluorescence of PicoGreen in the presence of varying concentrations of calf thymus DNA*

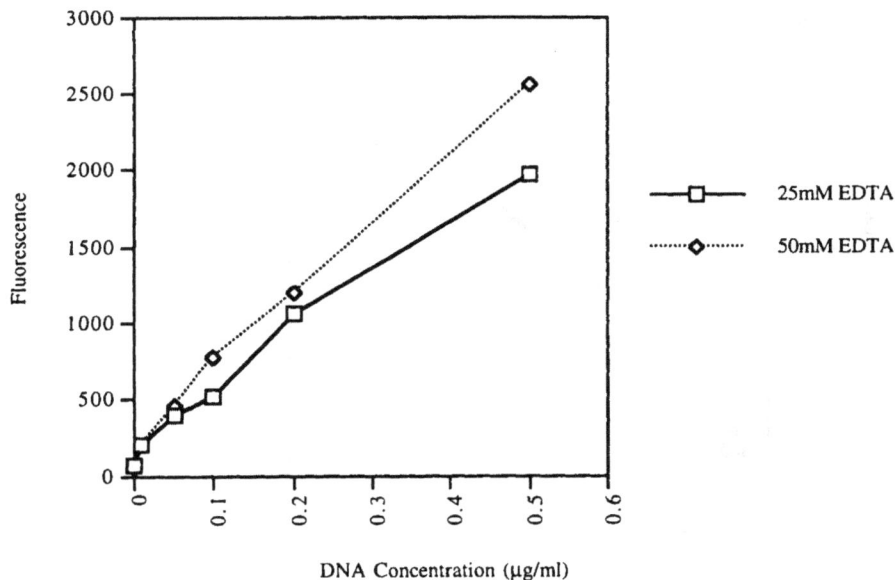

Figure 4.2 *Effect of EDTA on the fluorescence of PicoGreen in samples containing MgCl₂ (50 mM) and varying concentrations of calf thymus DNA*

as Na^+ and K^+ have only a modest effect on fluorescence and thus may be tolerated in quite high concentrations.[12]

The analysis of increasingly degraded DNA in the presence of PicoGreen tends to give a reduction in fluorescence and thus leads to an underestimation of the determined DNA concentration (Figure 4.3).

Finally, reagents such as sodium dodecyl sulfate and phenol also have a detrimental effect on fluorescence and steps must be taken to ensure these reagents are removed from samples before concentration measurements are made.[12] A summary of an investigation at the LGC into the effects of various sample contaminants is given in Table 4.1.

4.3.3.3 Hoechst 33258

Hoechst 33258 is another dye which exhibits fluorescent enhancement on intercalation with DNA. The dye is excited at 365 nm and emits at around 458 nm when complexed with dsDNA. Hoechst 33258 shows similar specificity to PicoGreen in that it binds dsDNA (and to a lesser extent ssDNA) in preference to RNA and is unaffected by the presence of proteins. Again, conditions under which fluorescence measurements are made are important. Divalent metal ions such as Mg^{2+} (>10 mM) should be avoided as they reduce the fluorescent enhancement observed. However, monovalent metal ions, such as Na^+, have very little effect on fluorescence.[13] A disadvantage of Hoechst 33258 is that it

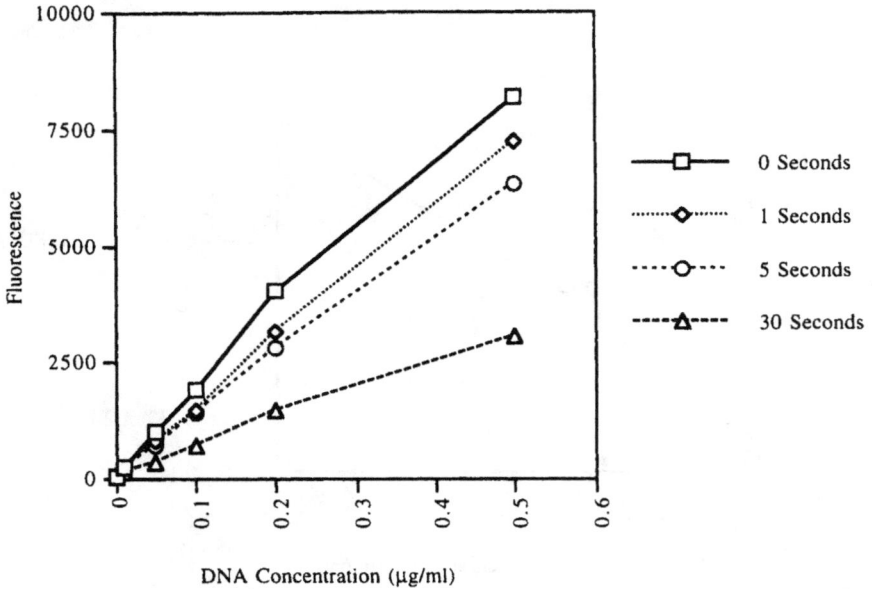

Figure 4.3 *Effect of DNA degradation on the fluorescence of PicoGreen. Calf thymus DNA was degraded by sonicating for increasing periods of time as shown*

Table 4.1 *Summary of the effect of various contaminants on the fluorescence of PicoGreen*

Contaminant (concentration)	Fluorescence change (compared to control without contaminant)
Bovine serum albumin (2 mg/ml)	−16%
Phenol (0.1% v/v)	+13%
SDS (0.1% w/v)	−90%
$MgCl_2$ (100 mM)	−72%
$CaCl_2$ (100 mM)	−74%
$ZnCl_2$ (5 mM)	−43%
NaCl (100 mM)	−12%
KCl (100 mM)	−11%

binds specifically to A–T regions. This can cause problems when the dsDNA used to plot the calibration curve is different from the test samples. For example, if an A–T rich dsDNA standard is used to plot the calibration curve to determine the concentration of DNA samples from various sources, which are either A–T or G–C rich, then the DNA samples which were G–C rich would have an artificially low measured concentration. Thus when measuring genomic DNA concentration with Hoechst 33258 it may be advisable to plot calibration curves using genomic DNA from a related species to the test DNA.

4.4 Summary

Determination of genomic DNA concentration has become more important as analytical techniques have become more sensitive. The concentration of starting material required to give optimum results has become well defined and in some cases covers an increasingly narrow range. Accuracy of determinations are difficult to assess due to the lack of suitable reference materials; reproducibility of determinations, however, can be improved by the employment of good quality samples, trained staff, calibrated equipment, suitable in-house standards, and use of methodologies within their limitations (in particular, minimum and maximum detection limits where linear relationships are maintained).

4.5 References

1. Hunt, D. J., Parkes, H. C. and Lumley, I. D. 1997. Identification of the species of origin of raw and cooked meat products using oligonucleotide probes. *Food Chem.* **60**: 437–442.
2. Waye, J. S., Michaud, D., Bowen, J. H. and Fourney, R. M. 1991. Sensitive and specific quantification of human genomic deoxyribonucleic acid (DNA) in forensic science specimens: casework examples. *J. Forensic Sci.* **36**: 1198–1203.
3. Andersen, J. 1998. Quantification of DNA by slot-blot analysis. In: Forensic DNA Profiling Protocols (eds. Lincoln, P. J. and Thomson, J.), pp. 33–38. Humana Press, Totowa, NJ.
4. Winfrey, M. R., Rott, M. A. and Wortman, A. T. 1997. Unraveling DNA — Molecular Biology for the Laboratory, pp. 56–59, 73–80, 282–284. Prentice-Hall, Englewood Cliffs, NJ.
5. Killeen, A. 1997. Quantification of nucleic acids. *Clin. Lab. Med.* **17**: 1–19.
6. Sambrook, J., Fritsch, E. F. and Maniatis, T. 1989. Molecular Cloning — A Laboratory Manual, vol. 3, p. E5. Cold Spring Harbor, New York.
7. Wilfinger, W. W. 1997. Effect of pH and ionic strength on the spectrophotometric assessment of nucleic acid purity. *BioTechniques* **22**: 474–481.
8. Kallansrud, G. and Ward, B. 1996. A comparison of measured and calculated single- and double-stranded oligodeoxynucleotide extinction coefficients. *Anal. Biochem.* **236**: 134–138.
9. Manchester, K. L. 1995. Value of A260/A280 ratios for measurement of purity of nucleic acids. *BioTechniques* **19**: 208–210.
10. Ahn, S. J., Costa, J. and Emanuel, R. 1996. PicoGreen quantitation of DNA: effective evaluation of samples pre- and post-PCR. *Nucleic Acids Res.* **24**: 2623–2625.
11. Sambrook, J., Fritsch, E. F. and Maniatis, T. 1989. Molecular Cloning — A Laboratory Manual, vol. 3, pp. E.8–E.9. Cold Spring Harbor, New York.
12. Singer, V. L., Jones, L. J., Yue, S. T. and Haugland, R. P. 1997. Characterisation of PicoGreen reagent and development of a fluorescence-based solution assay. *Anal. Biochem.* **249**: 228–238.
13. Labarca, C. and Paigen, K. 1980. A simple, rapid and sensitive DNA assay procedure. *Anal. Biochem.* **102**: 344–352.

CHAPTER 5

PCR: Factors Affecting Reliability and Validity

DAVID McDOWELL

5.1 Introduction to the Polymerase Chain Reaction

The polymerase chain reaction (PCR)[1,2] is an immensely powerful genetic amplification technique offering high levels of sensitivity and specificity in bioanalysis. However, in order to achieve maximal performance, great care has to be exercised at all stages of the procedure. There are many variations on PCR which will not be covered here. Instead, generic aspects of PCR will be discussed to draw attention to some of the potential pitfalls in order that basic amplifications can be carried out in the most effective way and valid data produced.

Briefly, the PCR reaction is initiated from two short synthetic pieces of DNA, known as primers, which are homologous to opposing ends of a selected sequence and delineate the region to be amplified. The specificity of the reaction is dictated by the uniqueness of the priming sequence within the template DNA used in the assay, and control of the stringency under which the primers are allowed to interact with target as opposed to non-target sequences. The sensitivity of the reaction results from sequential rounds of amplification, controlled by a thermal cycler, in which the products of one cycle can potentially act as targets in the next cycle. The process is repeated in the region of 30 times, allowing the amount of the selected target to be theoretically doubled at each cycle. In practice, the efficiency of the amplification approaches 100% for only a limited part of the process.

Eventually the amount of DNA produced in an amplification reaction will reach a maximum level known as the plateau. Whilst the number of PCR cycles before plateau is attained is dependent upon the amount of target at the start of the reaction, the level of plateau (copies of product) is independent of the initial target concentration. Consequently, the yield of DNA in reactions amplified to plateau cannot be used for quantification purposes. There are, however, ways of achieving quantification such as competitive PCR[3] and real time PCR[4-6] which are beyond the scope of the discussion here.

The exponential nature of the amplification process means that subtle

Table 5.1 *Selected features of basic PCR*

Benefits
- Sensitivity — Can detect very low numbers of target DNA. Theoretically, single copy number detection is possible under ideal conditions but substantially higher amounts of target may be required, depending upon the type of sample
- Specificity — Can be used to identify a specific genetic sequence accurately within a complex genetic background such as a whole genome

Potential Pitfalls
- Lack of precision due to block-to-block, run-to-run or tube-to-tube variation
- False negative results may arise as a consequence of inhibition or genetic polymorphism at selected priming sites
- False positive results due to inadvertent contamination incurred during PCR set up, perhaps due to the generation of aerosols

Requirements for Effective Use
- Good planning and experimental design
- Correct use of suitable controls (positive, negative and non-target)
- Correct use of thermal cycler
- Consideration of thermal profile and effect on results
- Good laboratory set-up/housekeeping, equipment calibration
- Thorough optimisation and validation

differences in amplification efficiency can lead to relatively large differences in product yield and results. Tube-to-tube variation of amplification efficiency can result from pipetting differences between reactions or variations in temperature between different positions within the thermal cycler block. Variations can also occur between different runs on the same thermal cycler, different 'identical' machines, different makes of machine or different batches of reagents. The generation of false positive results due to the presence of contaminating DNA poses an additional threat. There is therefore a requirement for both calibration of the thermal cycler and the use of suitable positive and negative controls in order to have confidence in the results obtained.

There are many excellent texts on PCR to which the reader should refer for further detail or more specialised applications.[7-10] Some of the features of basic PCR are given in Table 5.1.

5.2 An Illustrative Basic Amplification Protocol

In order to consider the potential pitfalls of PCR-based amplification, it is useful to have an illustrative protocol for discussion purposes. The following protocol can perhaps be viewed as typical and will serve as a starting point for such a discussion. Selection of the appropriate parameters for any given application should be based on careful optimisation of reaction components, the performance of the thermal cycler to be used and the specific question to be answered. Careful consideration and planning of all stages of the amplification process is required if a robust and reliable assay is to be performed.

Table 5.2 *An illustrative 50 μl PCR amplification*

10^5	Copies of template DNA
1 μM	Each primer
1.25 U	*Taq* DNA polymerase
10 mM	Tris-HCl pH 8.3
1.5 mM	$MgCl_2$
50 mM	KCl
200 μM	Each deoxynucleotide triphosphate (dNTP)
50 μl	Final volume with sterile distilled water

A standard amplification reaction as set out below might be 25–100 μl in volume, although volumes as small as 5 μl have been used successfully. The reaction would contain target DNA, deoxynucleotide triphosphates (dNTPs), a reaction buffer, a thermostable DNA polymerase such as *Taq* and oligonucleotide primers defining the start and end points of the final amplified product. A suitable *Taq* DNA polymerase buffer is generally supplied with the enzyme and may comprise a complete buffer and/or a magnesium-free buffer with separate $MgCl_2$ to allow optimisation work.

The amplification is normally performed in a small polypropylene tube placed in a thermal cycler programmed to achieve three selected temperatures in turn, known as a cycle. During each cycle, DNA denaturation, primer annealing and extension occur. About 25–35 cycles of amplification usually yields sufficient DNA over 2–4 hours for subsequent analysis by agarose gel electrophoresis, restriction endonuclease digestion, hybridisation or sequencing. A typical 50 μl PCR is illustrated in Table 5.2.

A typical thermal profile for a cycler using a simulated tube or equivalent to assess reaction temperature (see Section 5.3.3) is given in Table 5.3.

Such a profile gives an initial longer hold at 94 °C to ensure complete denaturation of DNA; this may be particularly relevant if the template is chromosomal in origin. In subsequent cycles, previously amplified product serves as template which denatures relatively easily. A final hold time is often used, ostensibly to ensure all product is full length, and is of limited value. There is, however, evidence that longer hold times increase the number of products

Table 5.3 *A typical thermal profile for PCR amplification*

94 °C	30 seconds	Hold
94 °C	30 seconds	
55 °C	30 seconds	30 cycles
72 °C	30 seconds	
72 °C	5 minutes	Hold

having an extra 3′ dA residue resulting from the template independent polymerase activity of *Taq*.[11] This may improve the efficiency of cloning in procedures dependent upon the dA overhang.

5.3 Critical Factors Affecting Amplification and Reproducibility

Almost every factor associated with PCR can potentially affect the performance of the amplification. In order to generate a robust and specific amplification protocol, a number of factors need to be optimised. Furthermore, if a protocol is to be effectively and repetitively used or transferred successfully between laboratories, great care needs to be taken in the way that such an optimised protocol is employed. The following sections describe some of the parameters affecting performance and reproducibility, which should be considered.

5.3.1 Reaction Components

5.3.1.1 Magnesium Chloride

PCR-based amplification utilises a thermostable DNA polymerase for DNA synthesis. DNA polymerases are magnesium dependent and therefore the major role of magnesium within the amplification mix is to serve as an enzyme co-factor. A number of factors influence the level of magnesium required for optimal reaction performance. A magnesium chloride concentration of 1.5 mM is suitable for many applications and is recommended as a good starting point. When considering the concentration of magnesium, it is important to remember that dNTPs bind magnesium with a stoichiometry of 1:1. In the reaction quoted (Section 5.2) the total dNTP concentration would be 0.8 mM and the level of free magnesium is therefore 0.7 mM.

Optimisation of free magnesium in the range of 0.5–2.5 mM is usually sufficient for most applications, although the use of substantially higher levels may be of benefit for certain types of sample. For example, in an assay for the direct detection of *Listeria monocytogenes*, milk was included directly in the amplification mixture. Increasing the total magnesium concentration from 1.5 mM to 6 mM was shown to raise the level at which successful amplification was possible from 5% to 20% milk content.[12] It should also be borne in mind that other reaction components, including template DNA, chelating agents such as EDTA, and proteins can also affect the amount of free magnesium. Different enzymes may also require different buffer conditions and it is therefore important to check the supplier's guidelines before using them for the first time.

As a guideline, too much free magnesium may reduce the specificity of the reaction whilst free magnesium levels of less than 0.5 mM may compromise the activity of the enzyme and reduce the yield of product obtained.

5.3.1.2   *Thermostable DNA Polymerases*

A number of thermostable DNA polymerases can be used for PCR, although for the majority of purposes the enzyme of choice is *Taq* polymerase. *Taq* is usually used at 1.25 units/50 μl reaction. This is an excess of enzyme for most purposes. The use of additional enzyme to improve product yield is unlikely to be successful and will increase the likelihood of generating non-specific products seen as artefactual bands and smearing on gel analysis. It is difficult to pipette accurately small volumes of enzyme stored in 50% glycerol owing to the viscosity of the solution. The use of master mixes for a set of reactions is therefore recommended as this can substantially reduce pipetting errors, thereby giving a greater uniformity amongst a group of amplification reactions by eliminating the accidental addition of too much or too little enzyme. When it is considered necessary to pipette small volumes of enzyme, it may be possible to dilute this in an appropriate buffer prior to use.

The recommended dilution buffer for Amplitaq® DNA polymerase is given in Table 5.4. The use of alternative enzymes is discussed in Section 5.3.5.

Table 5.4 *Dilution buffer for AmpliTaq DNA polymerase*

0.15% Nonidet P-40
0.15% Tween 20
0.1 mM EDTA
25 mM Tris-HCl (pH 8.3 at room temperature)

5.3.1.3   *Deoxynucleotide Triphosphates*

A dNTP concentration range of 50–200 mM for each dNTP is usually quoted as being optimal for most applications. The concentration can be optimised to maximise specificity, although the product yield may be compromised at the lowest levels. The use of high concentrations and unbalanced ratios of dNTPs are likely to compromise the fidelity of incorporation. Again it is important to consider the free magnesium concentration when adjusting the dNTP level.

5.3.1.4   *DNA Target*

It is also important not to add too much or too little DNA. Too little may give a negative result whilst too much may encourage the production of non-specific products seen as artefactual bands on gel analysis. As a guideline, 10^4–10^5 target copies usually results in a good amplification. In principle, lower levels can be used and even single copy detection is possible. When using template preparations containing very low levels of target DNA, additional precautions may be required such as replicate analyses. This could overcome any statistical variation inherent in removing small aliquots from a larger sample containing low levels of a target. Similarly, selecting suitable representative starting material for

Table 5.5 *Selected genome sizes*

1 μg pGEM® plasmid DNA = 2.85×10^{11} copies
1 μg lambda phage DNA = 1.9×10^{8} copies
1 μg human genomic DNA = 3.04×10^{5} copies

DNA extraction may also prove to be a sampling problem and needs to be considered (see Section 3.4.2). When considering the amount of DNA to be included, it is important to consider the genome size, as illustrated in Table 5.5.

In the majority of analytical situations, the amount and integrity of the DNA often limits the type of analysis that is possible. Where limited sample material is available, PCR is often the preferred technique. Where such samples are aged or processed, PCR is often the only choice as the sensitivity of the technique means that relatively few targets may need to be intact in order for analysis to be possible.

When amplifying from DNA which may be degraded, it is advisable to select as small a target as is feasible since these are more likely to have survived intact and are preferentially amplified to larger targets. This point is illustrated in Figure 5.1, where target DNA degraded to various degrees by sonication was added to PCRs containing a set amount of intact, high molecular weight DNA

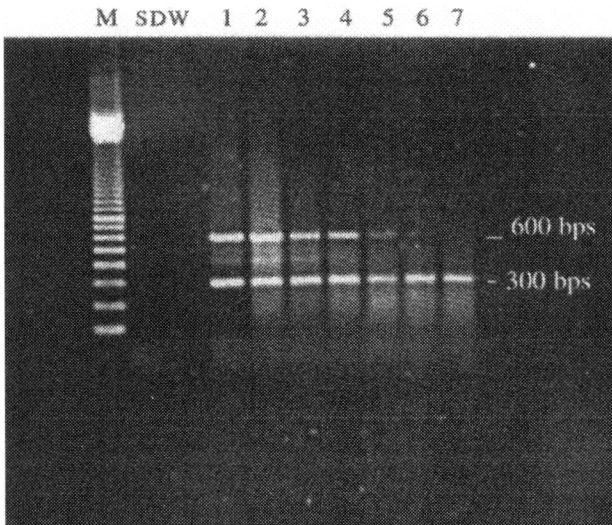

Figure 5.1 *The effect of using increasingly degraded DNA as a template for PCR with universal primers. Lane 1, amplification using HMW calf thymus DNA only; lanes 2–7, amplification using a mixture of DNA (70% degraded by sonicated + 30% HMW calf thymus). Sonication times were 0 s (lane 2), 10 s (lane 3), 30 s (lane 4), 50 s (lane 5), 70 s (lane 6), and 90 s (lane 7). M represents a 100 bp molecular weight marker and SDW represents a PCR negative control*

prior to amplification. It can be seen that the presence of increasingly more degraded DNA is progressively more inhibitory to amplification. This presumably results from the increasing numbers of interrupted targets and cross-hybridising fragments which may also serve as substrates for the DNA polymerase and therefore reduce amplification of the intended target. However, the persistent ability to amplify the smaller target (approximately 300 bp) demonstrates the value of selecting regions of 100–300 bp for successful amplification and analysis under such circumstances.

5.3.2   Primer Design and Target Selection

A number of guidelines exist for the design of primers for PCR. Whilst not following them does not necessarily guarantee failure, equally, following them does not guarantee success. Primer synthesis is now relatively cheap and it may well be cost effective to try several primer sets to achieve the desired result rather than spend large amounts of time attempting to optimise what may turn out to be a sub-optimal primer pair. Guidelines which may help to achieve success in primer design are given in Table 5.6.

A number of computer packages are also available, e.g. Oligo, Primers and PrimerSelect (DNAStar), which will assist in primer design but again do not guarantee success. When selecting a primer pair, it is also important to consider the size of product to be amplified. Whilst smaller targets will allow the use of shorter extension times and increase the speed of analysis, it may also be important to be able to distinguish between the desired product and any primer dimer formations when analysing results using standard electrophoresis. A target of approximately 100–200 bases is generally considered to offer a reasonable choice of sequence for the design of primers.

Larger target sizes may be required if the product is to encompass suitable restriction sites for subsequent analysis. Multiplex PCR may require a range of sizes to be targeted if identification is based on the use of electrophoretic separation. For particular purposes, it may be necessary to amplify substantial lengths of DNA. Long PCR, which is a modification of the standard amplification, has been used successfully to amplify targets in the region of tens of kilobases.[13]

Whilst the major aim of primer design is to select the target, the primers can often be modified for additional purposes. Extra sequences may be added at the

Table 5.6 *Guidelines for primer design*

- Select a primer size of 20–30 bases
- Balance the predicted melting temperatures of the primer pair
- Try to select a GC content of 40–60%
- Avoid complementarity of the primer pair, particularly at the 3' end
- Avoid runs of purines, pyrimidines or repetitive motifs
- Avoid regions possessing significant secondary structure

5′ end to incorporate restriction endonuclease cleavage sites for cloning purposes[14] or to serve as further priming sites of use in the construction of mimics in competitive PCR.[15] The design of primers with base changes to the original sequence can be used for site directed mutagenesis[16] or the introduction of promoter sequences to allow *in vitro* transcription.[17,18] A range of chemical modifications are also available for standard oligonucleotide synthesis which may be exploited in PCR for capture and detection purposes (e.g. addition of biotin, fluorophores and digoxigenin).

5.3.3 Thermal Cyclers: Factors Affecting Temperature Control

Technology has moved on dramatically since the early days of PCR when thermal cycling was achieved by manually moving reactions between different temperature water baths. There are now a range of automatic thermal cyclers on the market. Whilst many appear to be the same or at least very similar, differences between the machines and individual machine performance exist and can have profound effects on the end result. These need to be considered in order to ensure the effective use of PCR. Some thermal cyclers, which are designed for specific applications, such as those using glass capillaries as reaction vessels, have dramatically different specifications and reaction conditions and are not designed to be comparable with the range of more standard cyclers.

5.3.3.1 Temperature Profile

When considering the effect of the thermal cycler on PCR, it is important to understand what occurs during each cycle. There is a common misconception that discrete processes occur during three-temperature PCR, with DNA denaturation occurring at one temperature, primer annealing at a second and primer extension occurring at a third. A more realistic view suggests it is a dynamic process in which denaturation progresses to annealing and subsequently to extension before the cycle is repeated. Denaturation is likely to be complete as the set temperature is approached, particularly in later cycles when the substrate is predominantly amplified product as opposed to original target. Annealing starts as the set temperature is approached, becoming progressively faster as the optimal temperature is attained and potentially continues past the set annealing temperature until the temperature again becomes limiting. Once primer annealing has occurred, extension will follow potentially even before the set annealing and extension temperature has been attained since *Taq* is active over a wide temperature range, as shown in Table 5.7.

From Table 5.7 it may be seen that if erroneous priming occurs, extension can take place at relatively low temperatures if the DNA polymerase is in an active or accessible state. This can result in the generation of partial target containing a correct priming site for one of the primers. If such an event occurs for a second time with the second primer, a functional and incorrect target will have been generated containing both priming sites and which will compete with the

Table 5.7 *Amplitaq*® *extension rates*

Temperature	Extension Rate (bp/s)
70 °C	< 60
55 °C	24
37 °C	1.5
22 °C	0.25

intended target for amplification during each subsequent cycle. Steps to minimise such events occurring before thermal cycling is initiated, known as 'hot start' PCR, are described in Section 5.3.6. It is in the early cycles of amplification that false priming has the greatest potential for giving erroneous results since this is where they have the greatest potential for further amplification. Optimising the annealing temperature and time, along with the magnesium concentration, will limit the generation of erroneous products by limiting the potential for non-specific priming events.

In order to ensure the optimal result is obtained, it is important not only to prepare the reactions precisely but also to ensure that thermal cycling occurs correctly, with the annealing stage being considered the most critical. For a number of reasons, the temperature of the sample will lag behind the temperature of the block. The reaction has a certain volume and will take time to heat and cool and indeed to equilibrate. The reaction is generally in a plastic tube, which is a poor conductor of heat and which is itself heated by the block. If the thermal block is not kept clean, intimate contact will not be made with the tube and an insulating air gap will slow heat transfer still further. It is unlikely that such effects will be evenly distributed across the block, therefore creating a potential source of tube-to-tube variation.

The accuracy and precision of the temperature, and partly the time of annealing, will be limited by the equipment used. The uniformity of the block temperature and the speed with which the uniformity is achieved can be used to some extent to compare the quality of the machines. The effect of discrepancies in block uniformity is dramatically illustrated in Figure 5.2, which shows the same RAPD analysis performed on different makes of thermal cycler. A uniform pattern is visible for amplifications performed on one machine whilst highly variable profiles are generated with the other.

5.3.3.2 Temperature Control

The manner in which the thermal cycler is set up to achieve the correct temperature varies and this also needs to be considered if the machine is to be used correctly and in order to ensure thermal profiles applied to different machines are comparable.

- *Block control* — In machines using block control, the time the block is at a given temperature is measured. In such instances, hold times need to be

BRAND A   BRAND B

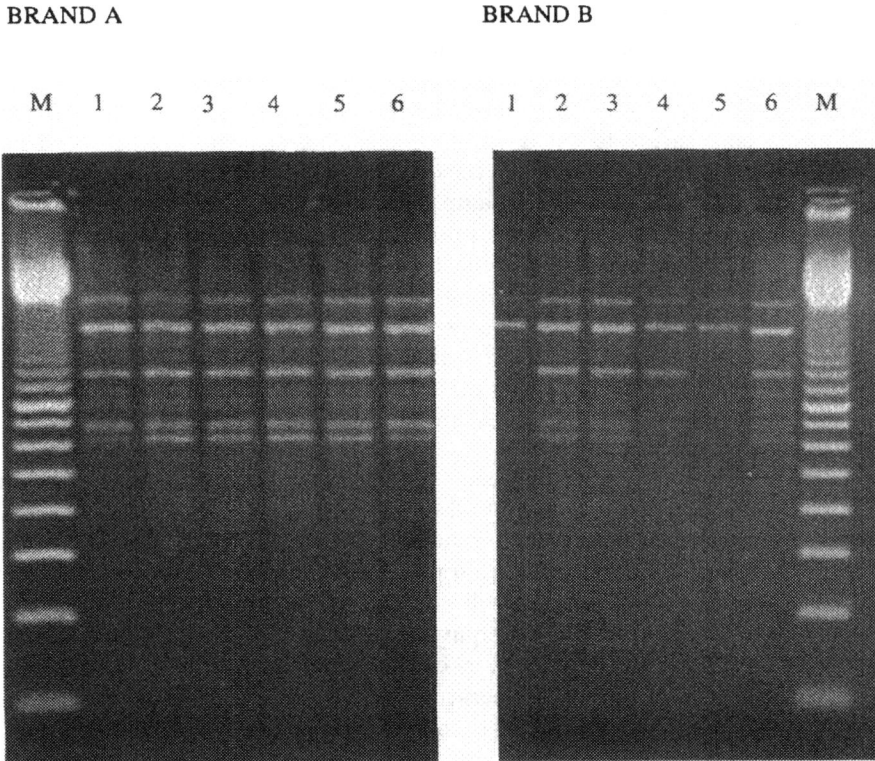

Figure 5.2 *RAPD profiles carried out using two different thermal cyclers. M = 100 bp molecular weight marker. Lanes 1–6 are repeats of the same reaction amplified at different positions within the block. Under the conditions used, cycler A gives consistent results, whereas cycler B shows considerable variation in the profile obtained*

relatively long in order to allow the sample to equilibrate with the block temperature. In such cases the use of short annealing times may improve specificity. This may result either from a reduction in the time available for mis-priming events to occur or from a failure of the reaction to attain the set annealing temperature in the time available and the consequent and inadvertent use of a higher and more specific annealing temperature. Similarly, the use of short denaturation times should be avoided as this may result in failure to achieve complete denaturation.

- *Reaction control* — Shorter hold times can be used where the hold time measured is the time the reaction (rather than the block) is predicted to be within a certain temperature window. Whilst such a window is often stated to be within 1 °C of the set temperature, this is not always the case and is a further hidden variable between different thermal cyclers. As the reaction temperature approaches the set temperature, two routes to achieving the final temperature may be taken.

1. The rate of temperature change decreases, allowing the sample temperature to catch up. The rate of temperature change and the temperature from which the new rate of change comes into effect will determine the time the sample remains within the given temperature hold window. This may result in differences in effective annealing, extension and denaturation times for different thermal cyclers, even when the same settings are used.
2. The block overshoots the set temperature for a given time before dropping back to the set temperature. The effect of this is to give a more rapid transition to the set temperature, which may again result in differences between machines.

There are, in addition, differences in the way machines assess sample temperature and this may compound differences between individual thermal profiles. Some have an internal simulated tube which is used to predict the reaction temperature at given or programmable volumes. Others use a reaction tube containing a temperature probe to estimate sample temperature. In all cases it is important to use an equivalent volume to that of the reaction. Failure to do so will result in an over- or under-estimate of the amount of time required to attain the correct temperature and an error in the time spent at the intended temperatures. The effect is compounded when the volume is significantly larger than that actually used and the machine uses a temperature overshoot to maximise the rate of transition to the set temperature. In such situations the reaction may overshoot the intended annealing temperature, potentially compromising the specificity of the amplification. Additionally, a higher than intended denaturation temperature may lead to increased heat inactivation of the *Taq* DNA polymerase, potentially reducing product yield, particularly where high sensitivity is required. A relatively small increase in denaturation temperature can have a comparatively large effect on the enzyme inactivation rate, as shown in Table 5.8. Similarly, it is important to use consistently either thick or thin walled reaction vessels as appropriate, as these will also affect the speed at which the correct sample tube temperature appears to be attained.

Table 5.8 *Amplitaq*® *inactivation rates*

Temperature	Enzyme Half Life (min)
97.5 °C	10
95 °C	40
92.5 °C	> 130

5.3.3.3 Ramp Rate

It is also possible in many instances to specify the speed of progression from one temperature to the next, otherwise known as the temperature ramp rate. For many amplification applications this is likely to have little effect. The exceptions may include short targets with short extension times where the extension occurs mainly during the ramp, and RAPD amplifications where significant amounts of priming may occur during the ramp to a low annealing temperature. Thermal cyclers having ambient or sub-ambient temperature capabilities will ramp at different rates irrespective of the set rate since ambient machines will be affected by room temperature which will vary from day to night and from season to season. Sub-ambient machines cooled by pumped water will be similarly affected by water temperature and flow rate. Depending on the way the machine is designed to function, this may lead to either a longer than predicted cycle time or failure to attain the set temperature before progressing to the next stage of the cycle.

To illustrate this point, Figure 5.3 shows a standard RAPD analysis performed on three different thermal cyclers under what was intended to be the same amplification conditions and thermal profiles. The first cycler (lane 1) had a sub-ambient temperature capability and used a simulated tube to assess sample temperature; the second cycler (lane 2) also had a sub-ambient capability but relied solely on block temperature and hold times to achieve the thermal profile; the final cycler did not have a sub-ambient capability and used air circulation to cool the block, with sample temperature being assessed by a separate reaction tube containing a temperature probe (lane 3). Distinct differences in RAPD profiles can readily be seen between these machines and

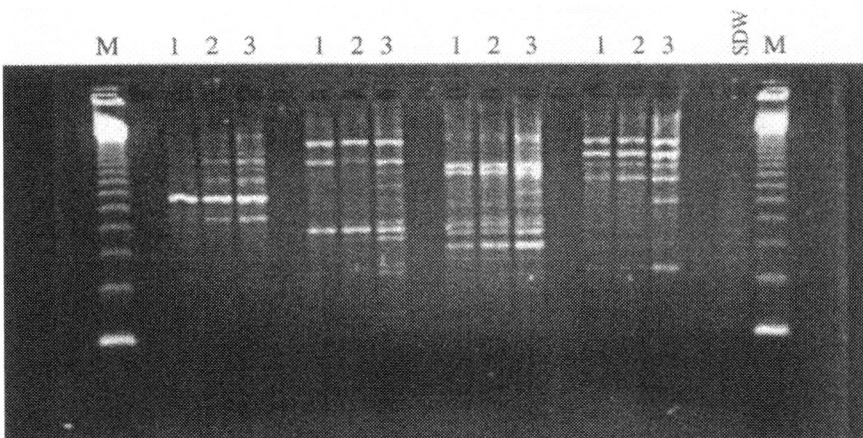

Figure 5.3 *RAPD profiles produced from three different thermal cyclers with four different 10 bp primers (A–D). Thermal cycler temperature control utilised an internal simulated tube to predict reaction temperature (lane 1), monitored block temperature (lane 2), and a thermal probe (lane 3) to assess the thermal profile. M represents 100 bp molecular weight marker*

illustrate the potential problems of moving protocols between different thermal cyclers. Measuring the time taken to complete a given number of cycles may be useful in estimating whether thermal profiles between machines are broadly similar, assuming the hold times at the set temperatures are correct.

5.3.3.4   Alternative Thermal Cyclers

Recent developments have sought to minimise amplification times. Such systems have used small reaction volumes in glass capillaries to give large surface area to volume ratios, which results in almost instantaneous temperature equilibration and minimal annealing and denaturation times. This, accompanied by ramp rates of 10–20 °C/s, made possible by the use of turbulent forced hot air systems to heat the sample, results in an amplification reaction completed in tens of minutes. Again with such technologies it is important to optimise the system correctly since the amount of time spent within a given temperature window is minimal if used at full speed and there is little chance of annealing occurring during the ramp to and from an incorrectly selected temperature.

5.3.3.5   Summary of Guidelines to Achieve Correct Thermal Profiles

Guidelines to achieve correct thermal profiles are summarised in Table 5.9.

Table 5.9 *Summary of guidelines to achieve correct thermal profiles*

Block temperature
• Check calibration — in-house/service agreement

Block uniformity
• Check uniformity — in-house/service agreement
• Keep block clean — see manufacturer's guidelines

Sample temperature
• Allow sufficient time for sample equilibration (block controlled cyclers)
• Consider the affect of ambient temperature according to machine
• Ensure simulated tube or temperature probe tube are comparable with reaction tubes. The use of different volumes or inappropriate thin/thick walled tubes will result in undershoot or overshoot of temperature (sample controlled cyclers)

General points
• Use cycler in accordance with manufacturer's guidelines
• When attempting to duplicate a thermal profile from a different machine, consider the way the cycler functions and adapt settings accordingly

---

5.3.4 Contamination in PCR: Good Housekeeping and Preventative Measures

The polymerase chain reaction is a very powerful genetic amplification tool. Theoretically able to detect a single DNA target by producing approximately 10^{12} copies of a selected sequence in only a few hours, the technique may easily be a victim of its own success. All the products of a PCR amplification are prime candidates for re-amplification and could potentially give rise to false positive results if not excluded from subsequent amplifications.

The risk of potential carry-over contamination should not be underestimated. By illustration, 10^{12} molecules of amplified product diluted in an Olympic sized swimming pool containing 2500 m^3 of water would result in four amplifiable molecules/μl.[19] Potentially as serious, although perhaps more surprising, is the threat posed by unnecessarily concentrated positive-control DNA stocks. For instance, a 2.5 kb plasmid control with a concentration of 1 μg/μl contains 3.6×10^{10} copies/0.1 μl. Aerosolisation of such a control may represent a severe contamination risk to other reactions. The centrifugation of DNA-containing tubes prior to opening can help to avoid splashing/aerosolisation, whilst keeping tubes closed and the frequent changing of gloves may reduce the risk of contamination. Sample-to-sample contamination may not necessarily represent as great a risk since a fairly gross contamination and/or extensive amplification would be required to give a false positive. However, such a theoretical consideration should not be used to endorse casual working practices as, in many cases, absolute confidence in the results is paramount.

A number of working practices and procedures exist for excluding unwanted DNA targets from amplification reactions and are discussed below. It is important that these are considered before work starts, since it can be difficult or expensive to resolve contamination problems after the event.

- Physical separation is one of the most widely cited means of controlling contamination and yet is perhaps the most difficult to set up and adhere to. Generally, a three-room system is suggested to serve as pre-PCR, PCR set-up and post-PCR areas. If space exists, there is also a very strong argument for using a fourth template addition room to ensure that PCR set-up and all associated reagents and consumables can be maintained in a DNA-free environment. In order to use such a system effectively, dedicated equipment and consumables are required for each area and the movement of DNA between sections by personnel limited by the use of area-restricted laboratory coats and pipettes, etc. Commercial, large-scale or routine services are recommended to use a 'one-way' progressive system for personnel involved in DNA extraction, PCR set-up and post-amplification analysis in order to enforce physical separation.
- Reagents and consumables should always be of the highest quality, sourced from reputable dealers and, where necessary, autoclaved before use. All PCR reagents are now commercially available or can be prepared from commercially available stocks. The preparation of relatively large batches of reagents stored as aliquots at $-20\,^{\circ}$C can help to minimise

quality control procedures and repeated freeze and thaw cycles as well as restricting any cross contamination.
- The use of positive displacement pipettes or filter tips when handling any DNA solution will prevent contamination of pipette barrels as a result of aerosolisation.

Other methods exist which seek to control contamination by destroying unwanted targets. Since prevention is better than cure, they are not recommended as a substitute for good working practice but may be of benefit where additional levels of confidence are required or where local working practices may of necessity compromise other control measures.

- By substituting dTTP with dUTP in amplification reactions, it is possible to degrade previous amplification products specifically by the use of uracil DNA glycosylase 'UDG'.[20] In practice, the PCR is set up with the inclusion of UDG. A preliminary incubation at 37 °C allows the destruction of any previously amplified product before a 10 min incubation at 95 °C is used to inactivate the UDG before amplification. Since UDG treatment and subsequent amplification is performed within a closed tube, there is no potential for further contamination. Whilst incomplete UDG inactivation may be a concern, residual activity has been shown to be insignificant during amplification and post-amplification degradation of products can be avoided by the use of a 72 °C soak, at which temperature the enzyme is inactivated.[21]
- Other methods for the inactivation of DNA are not specific for amplified products. The use of UV light has been recommended for the decontamination of surfaces[19] and for reagent solutions[22–24] and acts by dimerising neighbouring pyrimidines (CT, TT, TC, CC), thereby inhibiting polymerisation. In practice, the effect is reversible as irradiated targets exist in an equilibrium state, with the consequence that short PCR products may be relatively UV resistant, depending upon their sequence.[25] Achieving sufficient levels of exposure to UV is problematic where surfaces are not perpendicular to the light source, such as three-dimensional objects.
- Alternatives to UV involve chemical decontamination, which may be more applicable to three-dimensional surfaces, depending upon their chemical resistance. The effectiveness of hydrochloric acid and sodium hypochlorite (Clorox®) have been compared for this purpose.[26] In summary, a 1 min wash with 10% Clorox®, which was relatively non-corrosive, was sufficient to prevent PCR contamination, whilst a 5 min treatment with 2N HCl was found to be inefficient.

A different category of contamination originates from the polymerase used for the amplification. This is purified from a bacterial source and can contain contaminating DNA at levels estimated at 100 copies/unit of enzyme.[27] This may be a problem where there is homology between the primers selected and the

Table 5.10 *Precautions for minimising the potential for PCR contamination*

- Use dedicated areas and equipment
- Use filter tips or positive displacement pipettes
- Use DNA controls at sensible concentrations
- Use gloves to prevent cross contamination of samples
- Prepare large stocks of reagents and freeze in small aliquots replacing regularly
- Consider the use of UDG and dUTP to allow reactions to be purged of previously amplified product
- Consider the use of appropriate decontamination procedures for equipment and benches (UV, sodium hypochlorite)

contaminating DNA. This is particularly the case when amplifying a multicopy sequence such as the 16S rRNA genes using primers to highly conserved regions prior to sub-classification. In such instances it may be possible to eliminate contaminating DNA by the use of 8-methoxypsoralen and long-wavelength UV light.[24]

Precautions for minimising the potential for contamination of PCR amplifications are given in Table 5.10.

5.3.5 Polymerases: Features Affecting Selection and Use

The polymerase chain reaction was first demonstrated using the Klenow fragment of *E. coli* DNA polymerase.[1] The original process was very laborious since the DNA polymerase enzyme was not heat stable and consequently required the addition of fresh enzyme after each denaturation step. It was not until amplification was demonstrated using a thermostable DNA polymerase from *Thermus aquaticus*[2] that the true potential of the modern PCR process was realised. Whilst the *Thermus aquaticus* enzyme (*Taq*) remains the most widely used polymerase for PCR, a number of other enzymes are now in use for various applications. The factors affecting the choice of enzyme are summarised in Table 5.11.

Table 5.11 *Factors to be considered when selecting a thermostable DNA polymerase*

- What degree of thermal stability is required?
- What processivity is acceptable?
- Is a proof reading capability required (3′–5′ exonuclease), e.g. cloning and sequencing, long PCR or where internal regions my cause misincorporation and premature termination?
- Is a lack of proof reading capability required? A primer with a 3′ mismatch would not normally be expected to extend but would such a 3′–5′ exonuclease activity?
- What range of optimal magnesium concentration is required, e.g. multiplex PCR?
- Would the lack of a 5′–3′ exonuclease activity be beneficial, e.g. where more product/higher plateau is required such as multiplex PCR or RAPDs?
- Is an associated reverse transcriptase activity required: RT PCR?

The commonly used *Taq* enzyme does not have a proof reading 3'–5' exonuclease activity and can therefore introduce errors during *in vitro* DNA synthesis. For the majority of PCR-based detection and identification purposes, such errors are of little consequence and can be minimised by low and balanced dNTP concentrations. Work has shown that, for practical purposes, replication errors can be neglected if a large number of starting templates (e.g. 100 000 copies) are being used. For single locus analysis in single cells, the probability of false diagnosis due to such errors is of the order of 1%.[28] Where dNTP concentration is not limiting, incorporation rates may approach 150 nucleotides per second per enzyme molecule. By contrast, those factors which affect the specificity rather than the fidelity of PCR are generally of more consequence and therefore form part of the reason good optimisation is so important.

The two major factors which can be critical in optimising PCR performance are magnesium concentration and annealing temperature. Whilst these two variables are linked, the optimal level for different primer pairs may not be the same. The consequence of this may be difficulty in obtaining a suitable compromise when attempting to multiplex previously designed primer sets with different reaction optima. In such instances, one route to achieving a successful multiplex PCR has been to use a deletion derivative of *Taq* polymerase known as the Stoffel fragment,[29] which works under a broader range of magnesium concentration which may aid optimisation. In addition, this polymerase lacks the 5'–3' proof reading activity of *Taq* and consequently may synthesise more PCR product before plateau is attained, thereby reducing the sensitivity problems sometimes associated with multiplex PCR. The use of the Stoffel fragment may also be preferred with dideoxynucleotide sequencing because of the lack of the 5'–3' exonuclease activity. Its greater thermal stability also allows higher denaturation temperatures, which may be required for the amplification of GC-rich templates.

Where it is necessary to maintain the correct internal sequence during amplification, for instance where cloning and sequencing is to follow, it may be important to use a high fidelity enzyme with a 3'–5' exonuclease activity such as ULTma[TM] DNA polymerase[30] derived from *Thermotoga maritima* or an enzyme blend containing *Taq* and a small amount of a proof reading enzyme such as *Pfu*[TM] (derived from *Pyrococcus furiosus*). *Pfu* exhibits 11–12-fold greater replication fidelity than *Taq* polymerase.[31] Where misincorporation occurs, the product remains truncated in the absence of a proof reading 3'–5' exonuclease activity. Such products are inhibitory to efficient amplification, an effect which is particularly pronounced in 'Long PCR'.[13]

The use of enzyme blends may have particular application in situations where localised high melting temperature (T_m) regions are inhibitory to PCR. We have shown such high T_m regions to cause the formation of truncated products characterised by a template-independent misincorporation at the final base. The effect of this may not be significant for many standard PCR analyses. However, in competitive PCR, where it is the relative yield of two products using the same primer pair that is important, any differences in the amplification rate can have a pronounced effect, and even result in the apparent failure of one of the targets

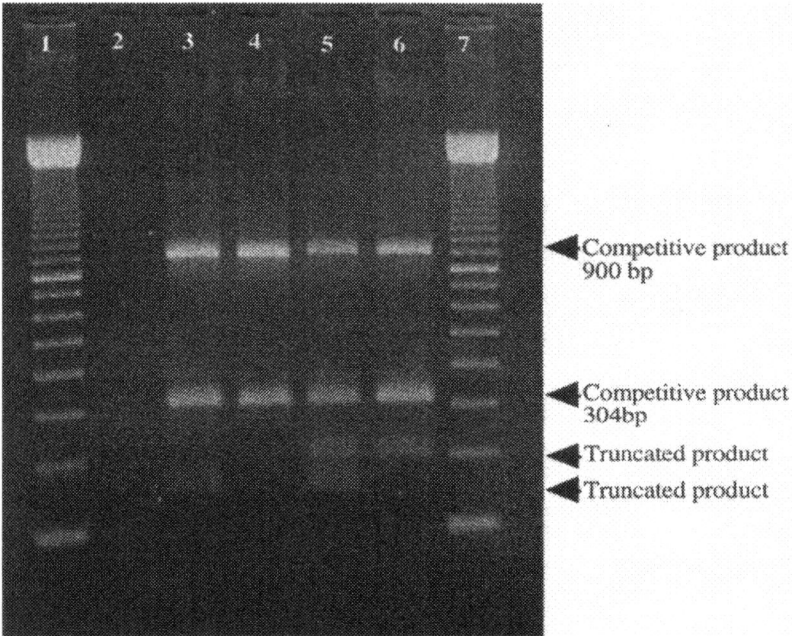

Figure 5.4 *Effect of the presence of betaine on the amplification of PCR products. Amplification was performed for 24 cycles before the removal of betaine followed by amplification for a further 16 cycles. Lane 1, 100 bp ladder; lane 2, negative control (water); lanes 3, 5 and 6, 900 bp DNA target (100 pg) + 53% GC 304 bp mimic (100 pg); lane 4, positive control (900 bp DNA target (100 pg) + 53% GC 304 bp mimic (100 pg), betaine not removed); lane 7, 100 bp ladder*

to amplify. We have shown the inhibitory effect of such regions to be reversible by the use of a small amount of the proof reading enzyme *Pfu* or the co-solute betaine.[32] By performing such a competitive reaction for a limited number of cycles in the presence of betaine and then continuing the amplification in the absence of betaine, the formation of truncated products can be readily visualised, as shown in Figure 5.4.

Difficulties in the amplification of long PCR targets are also thought to result from the erroneous incorporation of bases by *Taq* polymerase. Where long targets are being amplified, the potential for misincorporation is more pronounced and the chances of a full length product being produced is reduced.

A number of other thermostable DNA polymerases are available with a variety of biochemical characteristics. These include TLI/VENT™ from *Thermococcus litoralis*,[33] a number of polymerases from thermophilic *Bacillus* species including *Bst*™ from *B. stearothermophilus*,[34] SAC from *Sulpholobus acidcaldarius*,[35] TAC from *Thermoplasma acidophilum*,[36] TFL/TUB™ from *Thermus flavus*,[37] TRU from *Thermus rubra*,[38] TSP from *Thermotoga* spp.[39] and MTH from *Methanobacterium thermoautotrophicum*.[40]

Reverse transcriptase PCR (RT PCR) is another challenging area of PCR application where RNA rather than DNA is used as the starting material. Traditional reverse transcription can be carried out to generate a cDNA which can subsequently be used as a PCR target. Alternatively, a single enzyme reaction can be achieved by using the enzyme $rTth^{TM}$,[41] a recombinant form of the *Tth* polymerase from *Thermus thermophilus* which has both reverse transcriptase and DNA polymerase activities. Two strategies may be used with this enzyme. In the first, reverse transcription is carried out using $MnCl_2$, which is chelated out and replaced with $MgCl_2$ prior to PCR amplification. In the second, a single buffer system is used which minimises the potential for carry-over contamination as there is no requirement to open the tube between the reverse transcription and amplification stages.

5.3.6 Specificity

Many of the previously discussed points will affect the specificity of PCR-based amplifications. Ultimately, the specificity relies on the annealing of primers to correct targets and the ability of the polymerase to replicate the intervening region. A primer may, however, anneal to a region of DNA without it being a perfect match for that sequence. Depending upon the conditions used, it may do this with low efficiency. However, extension of that primer can have profound effects on the specificity of the reaction.

Erroneous targets generated by mis-priming events will amplify competitively with the intended target. The significance of their amplification will depend on the ratio of correct-to-incorrect target, their relative efficiency of amplification and the number of amplification rounds performed. The earlier such mis-priming events occur, the greater their likely significance. Because of this, many assays employ a technique known as hot start.[42] The aim of this is to prevent any primer annealing events, which occur during PCR set up, from extending until cycling conditions ensure that all primer template interactions are specific.

5.3.6.1 Hot Start PCR

Hot start PCR may be achieved in a number of ways and may or may not have a significant impact upon the result obtained, depending upon the nature of the assay being performed. The technique may be readily, if rather laboriously, performed by setting up reactions in two stages in which the reaction containing the primers and template is first denatured in the thermal cycler and held at 80–85 °C whilst the polymerase is added. Such a procedure should preclude any mis-priming and extension events. Care should be taken to dilute the *Taq* DNA polymerase in a suitable dilution buffer to ensure enzyme stability is maintained and to avoid pipetting very small volumes.

Manual hot start protocols can be very laborious if large numbers of samples are to be processed and can be an additional point during which contamination may occur. A number of automatic hot start approaches are now available.

These have used a number of methods to preclude erroneous extension prior to the first amplification cycle, including: layers of wax to separate portions of the reaction until denaturation has occurred; magnesium-containing wax beads that require melting of the bead before the magnesium-dependent polymerase becomes active; inactivation of the DNA polymerase using a *Taq* specific monoclonal antibody, which itself becomes heat inactivated during the first denaturation cycle, allowing the reaction to proceed; and, more recently, DNA polymerases requiring thermal activation. All such routes to hot start PCR should be effective in reducing non-specific amplification.

5.4 Concluding Comments

As can be seen from the preceding sections, there are very many variables which can affect the performance of PCR. All of these need to be considered and taken in context if a PCR-based procedure or analysis is to be used to its full potential. Whilst some variables can be excluded or at least minimised as sources of error, others can be optimised and exploited in order to achieve optimal results as described. Table 5.12 indicates those which should be considered first when optimising reaction conditions in order to improve PCR performance in terms of sensitivity, specificity and fidelity.

Table 5.12 highlights some of the actions that can improve PCR performance; it does not, of course, cover the whole story and is included as a summary to the text. Specificity and sensitivity are to some extent linked. For example, increasing the Mg^{2+} concentration may improve sensitivity but will eventually

Table 5.12 *Factors to consider in order to improve PCR performance in terms of sensitivity, specificity and fidelity*

Parameter	Adjustment	Effect on		
		Sensitivity	Specificity	Fidelity
Mg^{2+}	↑ ↓	↑ ↓	↓ ↑	
dNTP concentration	↓ ↑			↑ ↓
Annealing temperature	↑ ↓	↓	↑ ↓	
Cycle number	↑ ↓	↑ ↓	↓ ↑	
Hot start	+ −	↑ ↓	↑ ↓	
Proof reading enzyme	+ −			↑ ↓

↑ increase. ↓ decrease, + include, − don't include

compromise specificity. Since non-specific products will compete with the intended target for amplification, a loss in sensitivity will result. Careful optimisation of individual factors is therefore recommended in order to obtain optimal results. Additionally, whilst PCR enhancers may be used successfully to improve performance as discussed in Chapter 6, there may be instances where the best course of action is to redesign the primers since there is no guarantee that any primer pair will work in practice.

In order to have confidence in the results obtained from any PCR-based analysis, it is important to incorporate appropriate controls and to use them wisely and routinely. There is little value in using a series of negative controls to demonstrate a lack of contamination when those reactions are set up and capped before any chance of contamination occurs. When designing controls, bear in mind the questions which they are intended to answer. Negative controls need to be a worst case scenario or at least give a fair representation of events occurring during the analysis. This will result in a high degree of confidence that any positive results are not the result of contamination. Positive controls need to represent a snapshot in time by demonstrating that both reaction and thermal cycling conditions were achieved correctly. Controls should not have such a high template copy number as to be unrepresentative of the samples to which comparison is to be drawn. High template copy number controls are likely to be more tolerant of certain types of error. It should be the intention that both sample and control assays should pass or fail in tandem.

We have sought to illustrate here some of the pitfalls of PCR and some of the assumptions which may be made in error. When trying to repeat work carried out elsewhere, it is prudent to look not only at the description of the assay but also at the way in which it was carried out since this may indicate if, and how, the intended and achieved assays varied. The use of standardised protocols and procedures from which ambiguities have been eliminated can go a long way in achieving optimal PCR results and need to be encouraged. There appears, however, less hope at this stage of minimising the final two variables in the PCR process: the sample and the user. At least part of that variability is in your hands!

5.5 References

1. Saiki, R. K., Scharf, S., Faloona, F., Mullis, K. B., Horn, G. T., Erlich, H. A. and Arnheim, N. 1985. Enzymatic amplification of β-globin genomic sequences and restriction site analysis for diagnosis of sickle cell anemia. *Science* **230**: 1350–1354.
2. Saiki, R. K., Gelfand, D. H., Stoffe, S., Scharf, S. J., Higuchi, R., Horn, G. T., Mullis, K. B. and Erlich, H. A. 1988. Primer-directed enzymatic amplification of DNA with thermostable DNA polymerase. *Science* **239**: 487–491.
3. Wang, A. M., Doyle, M. V. and Mark, D. F. 1989 Quantitation of mRNA by the polymerase chain reaction. *Proc. Natl. Acad. Sci. USA* **86**: 9717–9721.
4. Higuchi, R., Dollinger, G., Walsh, P. S. and Griffith, R. 1992. Simultaneous amplification and detection of specific DNA sequences. *Bio/Technology* **10**: 413–417.
5. Higuchi, R., Fockler, C., Dollinger, G. and Watson, R. 1993. Kinetic PCR analysis; real-time monitoring of DNA amplification reactions. *Bio/Technology* **11**: 1026–1030.

6. Ishiguro, T., Saitch, J., Yawata, H., Yamagishi, H., Iwasaki, S. and Mitoma, Y. 1995. Homogeneous quantitative assay of hepatitis C virus RNA by polymerase chain reaction in the presence of a fluorescent intercalater. *Anal. Biochem.* **229**: 207–213.

7. Newton, C. R. and Graham, A. 1994. PCR. Bios Scientific Publishers, Oxford.

8. Griffin, H. G and Griffin, A. M. 1994. PCR Technology. CRC Press, Boca Raton, FL.

9. Mullis, K. B., Ferre, F. and Gibbs, R. A. 1994. The Polymerase Chain Reaction. Birkhauser, Boston.

10. Innis, M. A., Gelfand, D. H. and Sninsky, J. J. 1995. PCR Strategies. Academic Press, San Diego, CA.

11. Hu, G. 1993. DNA polymerase controlled addition of non-templated extra nucleotides to the 3' end of a DNA fragment. *DNA Cell Biol.* **12**: 763–770.

12. Bickley, J., Short, J. K., McDowell, D. G. and Parkes, H. C. 1996. Polymerase chain reaction (PCR) detection of *Listeria monocytogenes* in diluted milk and reversal of PCR inhibition caused by calcium ions. *Lett. Appl. Microbiol.* **22**: 153–158.

13. Cheng, S., Chang, S.-Y., Gravitt, P. and Respess, R. 1994. Long PCR. *Nature* **369**: 684–685.

14. Scharf, S. J., Horn, G. T. and Erlich, H. H. 1986. Direct cloning and sequence analysis of enzymatically amplified genomic sequences. *Science* **233**: 1076–1078.

15. Siebert, P. D. and Larrick, J. W. 1993. PCR MIMICS: competitive DNA fragments for use as internal standards in quantitative PCR. *BioTechniques* **14**: 244–249.

16. Landt, O., Grunert, H.-P. and Hahn, U. 1990. A general method for rapid site-directed mutagenesis using polymerase chain reaction. *Gene* **96**: 125–128.

17. Stoflet, E. S., Koeberl, P. A., Sarkar, D. D. and Sommer, S. S. 1988. Genomic amplification with transcript sequencing. *Science* **239**: 491–495.

18. Kain, K. C., Orlandi, P. A. and Lanar, D. E. 1991. Universal promoter for gene expression without cloning: expression PCR. *BioTechniques* **10**: 366–374.

19. Kwok, S. and Higuchi, R. 1989. Avoiding false positives with PCR. *Nature* **339**: 237–238.

20. Longo, M. C., Berniger, M. S. and Hartley, J. L. 1990. Use of uracil-DNA glycoslylase to control carry-over contamination in polymerase chain reactions. *Gene* **93**: 125–128.

21. Thornton, C. G., Hartley, J. L. and Rashtchian, A. 1992. Utilising uracil-DNA glycosylase to control carryover contamination in PCR: characterisation of residual UDG activity following thermal cycling. *BioTechniques* **13**: 180–183.

22. Sarkar, G. and Sommer, S. S. 1991. Removal of DNA contamination in polymerase chain reaction reagents by ultraviolet irradiation. *Methods Enzymol.* **218**: 381–388.

23. Isaacs, S. T., Tessman, J. W., Metchette, K. C., Hearst, J. E. and Cimono, G. D. 1991. Post-PCR sterilization development and application to an HIV-1 diagnostic assay. *Nucleic Acids Res.* **19**: 109–116.

24. Meier, A., Persing, D. H., Finken, M. and Böttger, E. C. 1993. Elimination of contaminating DNA within polymerase chain reaction reagents: implications for a general approach to detection of uncultured pathogens. *J. Clin. Microbiol.* **31**: 646–652.

25. Cone, R. W. and Fairfax, M. R. 1993. Protocol for ultraviolet irradiation of surfaces to reduce PCR contamination. *PCR Methods Applications* **3**: S15–S17.

26. Prince, A. M. and Andrus, L. 1992. PCR: How to kill unwanted DNA. *BioTechniques* **12**: 358–360.

27. Rand, V. H. and Houck, H. 1990. *Taq* polymerase contains bacterial DNA of unknown origin. *Mol. Cell. Probes* **4**: 445–450.

28. Krawczak, M., Reiss, J., Schmidtke, J. and Rösler, W. 1989. Polymerase chain reaction: replication errors and reliability of gene diagnosis. *Nucleic Acids Res.* **17**: 2197–2201.

29. Lawyer, F. C., Stoffel, S., Saiki, R. K., Chang, S.-Y., Landre, P. A., Abramson, R. D.

and Gelfand, D. H. 1993. High-level expression, purification, and enzymatic characterisation of full-length *Thermus aquaticus* DNA polymerase and a truncated form deficient in 5' to 3' exonuclease activity. *PCR Methods Applications* **2**: 275–287.

30. Lawyer, F. C. and Gelfand, D. H. 1992. The DNA polymerase I gene from the extreme thermophile *Thermotoga maritima*: identification, cloning, and expression of full-length and truncated forms in *Escherichia coli*. *Abstr. 92nd Gen. Meeting, Am. Soc. Microbiol.*, p. 200.

31. Lundberg, K. S., Shoemaker, D. D., Adams, M. W. W., Short, J. M., Sorge, J. A. and Mather, E. J. 1991. High fidelity amplification using a thermostable DNA polymerase isolated from *Pyrococcus furiosus*. *Gene* **108**: 1–6

32. McDowell, D. G., Burns, N. A. and Parkes, H. C. 1998. Localised sequence regions possessing high melting temperatures prevent the amplification of a DNA mimic in competitive PCR. *Nucleic Acids Res.* **26**: 3340–3347.

33. Neuner, A., Jannasch, H. W., Belkin, S. and Stettner, K. O. 1990. *Thermococcus litoralis* sp. nov.: A new species of extremely thermophilic marine archaebacteria. *Arch. Microbiol.* **153**: 205–207.

34. Kaboev, A. K., Luchkina, L. K., Akhmedov, A. T. and Bekker, M. L. 1981. Purification and properties of a deoxyribonucleic acid polymerase from *Bacillus stearothermophilus*. *J. Bacteriol.* **145**: 21–26.

35. Elie, C., Salhi, S., Rossignol, J. M., Forterre, P. and de Recondo, A. M. 1988. A DNA polymerase from a thermoacidophilic archaebacterium: evolutionary and technological interests. *Biochim. Biophys. Acta* **951**: 261–267.

36. Hamal, A., Forterre, P. and Elie, C. 1990. Purification and characterisation of a DNA polymerase from archaebacterium *Thermoplasma acidophilum*. *Eur. J. Biochem.* **190**: 517–521.

37. Kaledin, A. S., Slyusarenko, A. G. and Gorodetskii, S. L. 1981. Isolation and properties of a DNA polymerase from the extremely thermophilic bacterium *Thermus flavus*. *Biokhimiya* **46**: 1576–1584.

38. Kaledin, A. S., Slyusarenko, A. G. and Gorodetskii, S. L. 1982. Isolation and properties of a DNA polymerase from the extremely thermophilic bacterium *Thermus rubra*. *Biokhimiya* **47**: 1785–1791.

39. Simpson, H. D., Coolobear, T., Vermue, M. and Daniel, R. M. 1990. Purification and some properties of a thermostable DNA polymerase from a *Thermotoga* species. *Biochem. Cell. Biol.* **68**: 1292–1296.

40. Klimczak, L. J., Grummt, F. and Burger, K. J. 1986. Purification and characterisation of a DNA polymerase from the archaebacterium *Methanobacterium thermoautotrophicum*. *Biochemistry* **25**: 4850–4855.

41. Myers, T. W. and Gelfand, D. H. 1991. Reverse transcription and DNA amplification by a *Thermus thermophilus* DNA polymerase. *Biochemistry* **30**: 7661–7666.

42. D'Aquila, R. T., Bechtel, L. J., Videler, J. A., Eron, J. J., Gorczyca, P. and Kaplan, J. C. 1991. Maximizing sensitivity and specificity of PCR by preamplification heating. *Nucleic Acids Res.* **19**: 3749.

CHAPTER 6

Inhibitors and Enhancers of PCR

JANE BICKLEY AND DANIEL HOPKINS

6.1   Introduction

The polymerase chain reaction[1] can be inhibited or enhanced by many different substances arising from the native biological specimens or the method and reagents used to extract the DNA. A wide variety of biological specimens are used for PCR, including animal tissues and bodily fluids, bacterial samples, forensic and archaeological material and plant tissues. Moreover, many of these may be sourced from crude, environmental samples, e.g. foodstuffs, soil and sludge. Many of these crude preparations contain substances inhibitory to PCR, although the exact mechanisms of inhibition and whether there might be an antagonistic effect between individual components is not always known.

Assuming that only a fraction of any given sample or extraction solution is transferred to a PCR, only a few substances will reach concentrations where they are inhibitory on their own. Individual compounds may, however, become concentrated by the extraction method, e.g. by co-precipitation with the sample DNA. The inhibitory mode of action of some of the individual compounds might be linked with precipitation of the DNA, denaturation of DNA or the polymerase enzyme, binding of the necessary Mg^{2+} ions or adding an excess of Mg^{2+}.

In general, PCR workers have focused on sources of error in interpreting positive amplification results, particularly with regard to false positives from nucleic acid contamination. However, less attention has been given to false negative reactions. The exquisite sensitivity of the polymerase chain reaction makes this technique attractive as a clinical test for diagnosis of infectious diseases. Negative test results for infectious agents can, correctly or not, imply that the organism in question is not responsible for the patient's illness, influencing therapeutic decisions such as withholding antibiotic and antiviral drugs. False negative results may be caused by inhibitors such as glove powder, impure DNA, heparin or haemoglobin, for example.

The medical consequences of negative PCR results can be especially profound compared with action taken in response to negative cultures, since clinicians may assume that a negative amplification rules out the presence of even a few genomes, effectively eliminating that infectious agent from diagnosis. The use of appropriate positive controls can increase confidence in negative PCR results by ruling out amplification failure due to inhibition as a cause of lack of amplification products. For example, a non-competitive co-amplification of a control nucleic acid added to the reaction for the native nucleic acid of interest will confirm that reaction conditions are correct and whether inhibitors are present.[2] Unsuccessful amplification of the control and the target could infer the presence of PCR inhibitors and therefore a false negative result, whereas the successful amplification of the control but not the target infers a true negative result, whilst the successful amplification of both implies a true positive.

Many researchers have also reported substances that can be added to a PCR assay to increase the sensitivity or specificity. Although the mode of action has not been determined for these factors, many possible explanations have been proposed. The major drawback of all of these substances is that no single one can be employed in any particular PCR with the guarantee of success. Many substances will enhance a PCR at a specific concentration, but the transfer of these conditions to a different PCR may or may not result in an effect. For example, the addition of 50 mM ammonium chloride, ammonium acetate or sodium chloride (all common reagents used for DNA precipitation) to a *Taq* DNA polymerase activity assay results in mild inhibition, no effect or slight stimulation, respectively.[3] This very marked variation in the way in which chemicals and reagents affect PCR amplification gives an indication of the potential problems to be aware of when applying this technology to the vast range of different samples for which it can be used.

It is intended that this chapter should provide a broad overview of the many varied applications of the polymerase chain reaction in which either it is adversely affected by inhibition or its efficiency improved by enhancers. It is not possible to include an exhaustive list of every inhibitor and enhancer, as the literature on the subject is ever increasing. However, it is hoped that the most common occurrences have been addressed, and an indication given of the effect on the PCR of different concentrations of various compounds, to enable the reader to assess the likely outcome of including a particular compound in a PCR assay. A representative range of examples is also included to illustrate the potential inhibitory problems in many different areas of PCR application, including medical, forensic, food and agricultural usage.

6.2 Factors that Inhibit or Enhance PCR

The aim of this section is to provide a wide range of examples, obtained from the literature, of inhibitors and enhancers of PCR that are likely to be encountered in the fields of molecular biology, medicine, forensics, food and agriculture. Whether the PCR is suitable for a certain type of analysis may depend on the nature of the sample, particularly with respect to the presence of inhibitory components that may interfere with amplification, and hence affect the robustness of the assay. Sample variation may mean that different levels of an inhibitory component may be present, impacting on the reproducibility of a particular PCR test. The inclusion of either inhibitors or enhancers, whether intentional or not, in the PCR will also impinge on many other aspects of the reaction, including sensitivity, specificity and product yield. Inhibitors will clearly cause a reduction in one or all of these features, whilst enhancers are included to improve these aspects of amplification.

6.2.1 Protocol for Assessing the Effects of PCR Inhibitors and Enhancers

The following section describes preliminary work carried out at LGC to assess the effect of a number of inhibitors and enhancers on PCR. In order to gain further insight into the level of reduction or improvement in amplification, yields have been enumerated and inhibitory or enhancing concentrations reported. Our aim was to establish a model system against which a wide range of compounds could be tested. The use of a standard, commercial source of DNA template and the consistent use of reagents and buffers added confidence to the belief that variations in amplification could be attributed to the presence or absence of each test compound at a known concentration. Such an approach would be recommended to any worker attempting to establish a robust and reliable PCR system for application in the presence of potentially problematic samples, especially when they are known to contain putative inhibitors.

LGC protocol for assessing the effects of PCR inhibitors and enhancers (see Table 6.1)
1. PCR (50 μl reaction):
- 10 × buffer (Pharmacia)*   5 μl
- Universal primer 1 (20 μM)   2.5 μl
- Universal primer 2 (20 μM)   2.5 μl
- dNTPs (Pharmacia, 100 mM)   8 μl
- *Taq* DNA polymerase (Pharmacia, 5000 units/ml)   0.2 μl
- Sterile distilled water   17.8 μl
- Inhibitor/enhancer   5 μl
- DNA (*Saccharomyces cerevisiae* 15.4 ng/ml, Sigma)   9 μl

(*Pharmacia 10 × buffer contains 500 mM KCl, 15 mM MgCl$_2$, 100 mM Tris-HCl)

Table 6.1 *PCR conditions for test reaction*

Set temperature	Hold time	Cycles
94 °C	5 minutes	1 cycle
94 °C	1 minute	
50 °C	30 seconds	30 cycles
72 °C	2 minutes	
72 °C	7 minutes	1 cycle
4 °C	∞	hold

2. *PCR product analysis:*

PCR products were electrophoresed on 1% agarose gels containing ethidium bromide, and viewed on a UV light box. Gels were then scanned using a BioImager (MilliGen/Biosearch running under Solaris 2.x.).

Tables 6.2 and 6.3 summarise the data generated for the inhibitors and enhancers tested so far with our model system. Illustrative examples are given in Figures 6.1 to 6.4, which include both electrophoretic and graphical representa-

Table 6.2 *Inhibitors of PCR tested at LGC*

Inhibitor	Range tested	Inhibitory concentration
Tryptone soya broth (Oxoid)	0.1–10 µg/ml	>2.5 µg/ml
Potato starch (Sigma)	0.1–10 µg/ml	>2.5 µg/ml
Tartrazine (Sigma)	0.1–10 mM	>1.0 mM
Tannic acid/tannin (Sigma)	0.1–10 µg/ml	<0.1 µg/ml
Sodium chloride (Fisher)	1–50 mM	>25 mM
Calcium chloride (Fisons)	0.1–10 mM	>1.0 mM
Zinc chloride (BDH)	0.001–10 mM	>0.001 mM
Ethidium bromide (Sigma)	0.001–10 mM	>0.1 mM

Table 6.3 *Enhancers of PCR tested at LGC*

Enhancer	Range tested	Optimum enhancer concentration	Change (from control)	Inhibition (if any)
Betaine (Sigma)	0.05–5 µg/ml	1.0 µg/ml	+185%	none at 5.0 µg/ml
Bovine serum albumin (Sigma)	0.5–5 µg/ml	1.0 µg/ml	+89%	none at 5.0 µg/ml
Dimethyl sulfoxide (Sigma)	0.05–5%	5%	+55%	none
Formamide (Amresco)	0.05–5%	0.1%	+28%	total inhibition at 5%
Glycerol (Sigma)	0.05–5%	0.05%	+144%	none at 5%
$MgCl_2$ (Sigma)	0.5–10 mM	5 mM	+2793%	inhibition at 0.5 mM
Polyvinyl-pyrrolidone (Sigma)	0.05–5 µg/ml	2 µg/ml	+59%	none at 5.0 µg/ml
Tetramethyl-ammonium chloride (Sigma)	0.05–5 mM	0.05 mM	+20%	>0.5 mM results in some inhibition
Triton X-100 (Sigma)	0.05–5%	0.05%	+137%	none at 5%

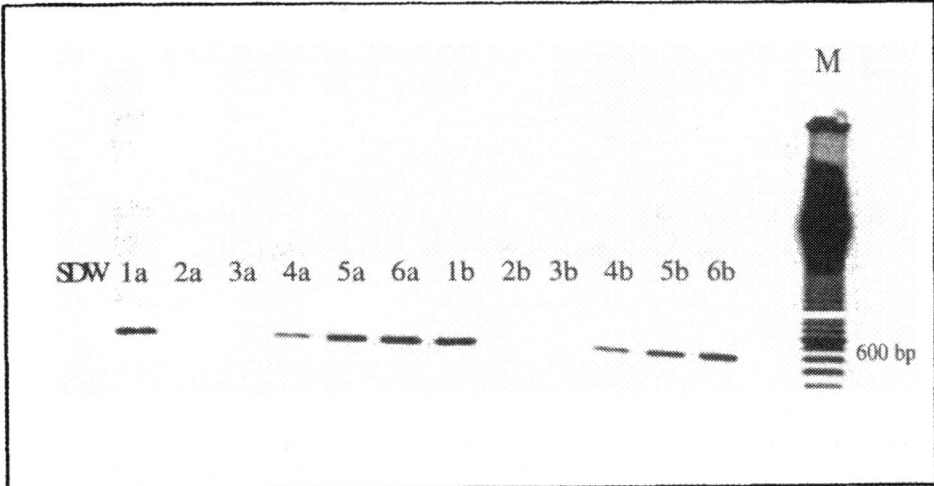

Figure 6.1 *A gel image showing the inhibiton of a PCR by various concentrations (in duplicate) of tryptone soya broth (TSB). Control reaction without the inhibitor, lanes 1a and 1b; 30 μg/ml TSB, lanes 2a and 2b; 15 μg/ml TSB, lanes 3a and 3b; 3.0 μg/ml TSB, lanes 4a and 4b; 1.0 μg/ml TSB, lanes 5a and 5b; and 0.1 μg/ml TSB, lanes 6a and 6b. M represents a 100 bp molecular weight marker*

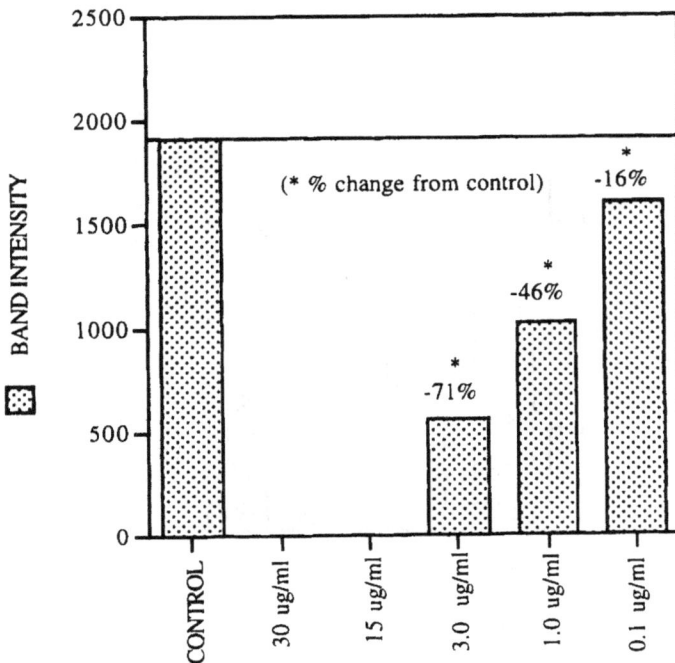

Figure 6.2 *A graph showing the inhibition of a PCR, in terms of band intensity, by various concentrations of tryptone soya broth*

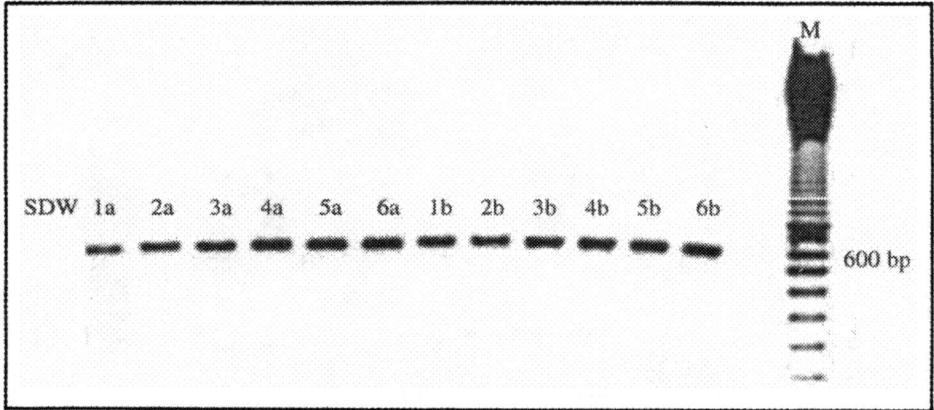

Figure 6.3   *A gel image showing the enhancement of a PCR by various concentrations of BSA (in duplicate). Control reaction without the enhancer, lanes 1a and 1b; 5.0 µg/ml BSA, lanes 2a and 2b; 2.5 µg/ml BSA, lanes 3a and 3b; 1.0 µg/ml BSA, lanes 4a and 4b; 0.1 µg/ml BSA, lanes 5a and 5b; and 0.05 µg/ml BSA, lanes 6a and 6b. M represents a 100 bp molecular weight marker*

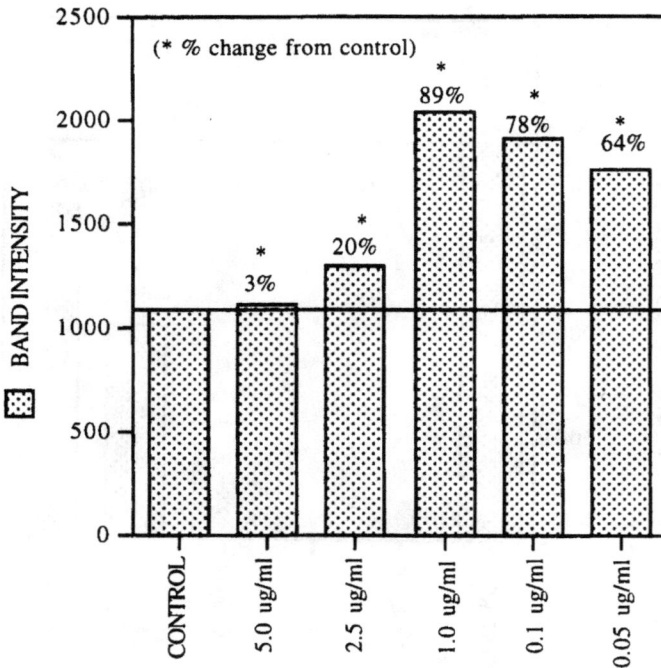

Figure 6.4   *A graph showing the enhancement of a PCR, in terms of band intensity, by various concentrations of BSA*

Table 6.4 *Concentration at which ionic (I) and non-ionic (N) detergents inhibit PCR*[5]

Detergent	Type	Inhibitory concentration
Na-sarkosyl	I	<0.02%
Na-deoxycholate	I	<0.06%
SDS	I	<0.01%
Nonidet P-40	N	>5%
Triton X-100	N	>5%
Tween	N	>5%
N-Octylglucoside	N	<0.4%

tions of typical results obtained. Figures 6.1 and 6.2 show the results for the inhibitory effects of tryptone soya broth, whilst Figures 6.3 and 6.4 demonstrate how the PCR can be enhanced with the addition of bovine serum albumin (BSA).

6.2.2 A Review of Inhibitors

6.2.2.1 Molecular Biological Reagents

DNA extraction methods can be quite different for diverse groups of biological specimens. In some instances, it may be possible to perform PCR directly on a clinical or food sample or a culture without prior DNA extraction. The initial denaturation step is often sufficient to cause cell lysis and release the target DNA, although in other cases a more rigorous short boiling step may be required. For many samples, various extraction procedures have been shown to increase the performance of PCR, although the possible carry-over of different inhibitory agents from the extraction procedure into the PCR reaction must also be considered. Re-extraction, ethanol precipitation and/or centrifugation may help to resolve this problem.

Extraction of DNA from different sources routinely uses detergents for cell lysis and denaturation. It has been shown that ionic detergents are far more inhibitory than non-ionic ones.[4] Non-ionic detergents (e.g. Triton X-100, Tween 20, Nonidet P40) are often used and these generally do not inhibit PCR at concentrations of up to 5% (Table 6.4). On the other hand, ionic detergents such as sodium dodecyl sulfate (SDS) can only be tolerated at extremely low concentrations (less than 0.01%) and therefore should be removed by phenol extraction and ethanol precipitation of the DNA before PCR. The inhibitory effects of low concentrations of SDS (e.g. 0.01%) can be reversed by the addition of certain non-ionic detergents (e.g. 0.5% Tween 20 and Nonidet P40).

The effect on PCR of a number of the compounds used in traditional extraction procedures has been investigated,[6] and is summarised in Table 6.5. Sodium hydroxide cannot be tolerated in concentrations of 8 mM and higher,

Table 6.5 *The concentrations of compounds commonly used in molecular biological procedures that are either tolerated or cause inhibition when added to the PCR*[6]

Compound	No inhibition	Inhibition
Phenol	0.1%	0.5%
Chloroform	5%	n.d.
Sodium dodecyl sulfate	0.005%	0.01%
Lithium dodecyl sulfate	0.005%	0.01%
Nonidet-P40	0.2%	2%
Tween 20	2%	10%
Triton X-100	1%	2%
Lysozyme	n.d.	0.5 mg/ml
Sodium hydroxide	5 mM	8 mM
Ammonium acetate	75 mM	150 mM
Ethanol	2.5%	5%
Isopropanol	0.5%	1%
Potassium acetate	0.02 M	0.2 M
Dithiothreitol	10 mM	n.d.
Bovine serum albumin	10 mg/ml	25 mg/ml
Ethylenediaminetetraacetic acid	0.1 mM	1mM
Ethylene glycol-bis(β-aminoethyl ether)-tetraacetic acid	0.1 mM	1mM
Spermidine	0.1 mM	1 mM
Proteinase K	0.5 mg/ml	n.d.
Bactotrypsin	5 μl/ml	n.d.
Ethidium bromide	0.1%	1%
Acetonitrile	2.5%	10%
Triethylammonium acetate	20 mM	75mM
Guanidinium isothiocyanate	20 mM	100 mM
Sarkosyl	0.01%	0.05%
Cetyltrimethylammonium bromide	0.001%	0.01%
Sucrose	5%	10%

n.d., not determined.

presumably because of pH-mediated denaturation of both the enzyme and of the double stranded DNA. The alcohols and poly(ethylene glycol) (PEG) may possibly precipitate the DNA, thereby stopping the reaction. There seems to be no difference between EDTA and EGTA as both inhibit the PCR at 1 mM but are acceptable at 0.1 mM, although only EDTA is known to chelate the essential Mg^{2+} ions.

Many detergent extraction procedures are performed in the presence of proteinase K, a protease that will digest denatured proteins. *Taq* DNA polymerase is susceptible to protease digestion, and so proteinase K must be removed or inactivated. Thermal denaturation at 95 °C followed by phenol extraction, which will also denature the proteinase K, is sufficient to achieve this. Residual traces of phenol, which inhibits PCR, can subsequently be removed by chloroform–isopentyl alcohol or ether extraction.

Low concentrations of urea, dimethyl sulfoxide (DMSO), dimethylforma-mide (DMF) or formamide have no effect on *Taq* polymerase's nucleotide incorporation activity. However, the presence of 10% DMSO in a 70 °C *Taq* polymerase activity assay inhibits DNA synthesis by 50%, and urea at 0.5 M completely inhibits a PCR assay. While several investigators have observed that inclusion of 10% DMSO facilitates certain PCR assays, it is not clear which parameters of PCR are affected. The presence of DMSO may affect the T_m of the primers, the thermal activity profile of *Taq* polymerase and/or the degree of product strand separation achieved at a particular denaturation temperature.

6.2.2.2 Gel Loading Dyes

In order to eliminate the extra pipetting steps required to prepare a PCR sample for gel analysis, coloured loading buffers that can be included directly in PCR have been investigated.[7] Many dyes, including the traditional DNA loading dyes bromophenol blue and xylene cyanol, completely inhibit PCR even at low concentrations. For screening purposes, dyes have been tested in the PCR at the lowest concentration that gives an intense enough colour to be seen easily. Metacresol purple, thymol blue, methylene blue, caramel colouring, india ink and red, green and blue food colours were all found to be inhibitory to amplification. Two dyes were found to be compatible with PCR, namely tartrazine and yellow food colouring, which contains cresol red dye. These PCR-compatible dyes each have unique properties. Neither cresol red nor tartrazine cause the dark shadows that are occasionally seen on ethidium-stained agarose gels with the traditional blue dyes. Tartrazine migrates quickly through the gel, approximately with the primers, and therefore is unlikely to be in the same location as the product bands. Cresol red migrates between bromophenol blue and xylene cyanol at approximately 300 bp in a 2% gel.

6.2.2.3 UV-damaged Mineral Oil

One established method to reduce false positives in PCR is the decontamination of pre-PCR reagents or laboratory equipment and surfaces by treatment with UV irradiation. Irradiation of DNA with UV light produces pyrimidine dimer adducts between adjacent pyrimidine bases, thymine bases in particular. The bases can therefore no longer form hydrogen-bonded base pairs and thus will not serve as a primer or a template for DNA synthesis.

It has been found that mineral oil repeatedly treated overnight with UV irradiation leads to inhibition of amplification reactions.[8] A typical PCR consists of an aqueous reaction mixture overlaid with mineral oil to prevent evaporation during incubation at elevated temperatures. To prove whether the inhibition of PCR was caused by UV-induced radicals interfering with *Taq* DNA polymerase, 8-hydroxyquinoline was tested, a substance well known as a very efficient trap for radicals. It was first confirmed that 8-hydroxyquinoline itself did not hamper PCR, and that PCR sensitivity was not affected if the 8-hydroxyquinoline was added prior to UV treatment of the mineral oil. If added to already UV-irradiated mineral oil, 8-hydroxyquinoline was shown to improve PCR sensitivity. Thus, during or after UV treatment, 8-hydroxyquinoline was able to trap the originating free radicals and prevent inhibition or degradation of *Taq* DNA polymerase. The same observation has been reported by another group of workers,[9] confirming that apparently inert components of PCR can affect the outcome of amplification.

6.2.2.4   Bodily Fluids

One of the most common sources of material for extraction of DNA is blood, whether it be for medical or forensic investigation. Heparin is a naturally occurring polysaccharide found primarily in the liver, lung and artery walls, and it is commonly used as an anticoagulant for blood samples. However, heparin has an inhibitory effect on the polymerase chain reaction and other enzyme-mediated reactions. The enzyme heparinase has been used successfully to eliminate heparin from DNA samples post-extraction, since the normal range of DNA extraction techniques fail to do so. The efficacy of different concentrations and grades of heparinase has therefore been examined for this purpose.[10] Other substances in blood, possibly porphyrin compounds, are also strong inhibitors of PCR and these can be eliminated from the DNA preparation by lysis of the red blood cells followed by centrifugation to pellet the DNA-containing white cells.

However, with bloodstains, a contaminant that interferes with PCR amplification is sometimes copurified with DNA following phenol/chloroform treatment and ethanol precipitation. It has been suggested that the contaminant is a heme compound, likely to result from the proteinase K digest of some heme blood protein complex, such as the complex of ferrous or ferric heme and serum albumin, known as hemalbumin or methemalbumin.[11]

Akane *et al.*[11] demonstrated characteristics of the contaminant in bloodstain extract by comparison with several heme compounds, suggesting that the contaminant was likely to be the heme–blood protein complex. The results of polyacrylamide gradient gel electrophoresis and intensity of the inhibition of PCR suggested that the ligand of the contaminant was a somewhat large molecule, resistant to proteolysis by proteinase K. The addition of BSA to the reaction mixture prevented the inhibition of PCR by the heme compounds, probably by binding to the heme. This showed that the inhibition was not due to the irreversible inactivation of the enzyme.

One of the most suitable methods for DNA extraction prior to PCR analysis has been organic solvent extraction followed by microconcentrator dialysis, with BSA added to the amplification reaction.[12] Another approach is to add hydrogen peroxide to specimens to decompose some inhibitors of PCR, such as the heme compound.[13] Ethanol precipitation should remove the hydrogen peroxide.

The susceptibility of Amplicor *Chlamydia trachomatis* PCR assays to inhibitory factors possibly present in cervical specimens has also been assessed.[14] Complete inhibition of the PCR was observed in 19% of cervical specimens.

Heat treatment (at 95 °C), freeze-thawing or 10-fold dilution of the samples reduced the initial inhibition to 9, 16 or 9%, respectively. A combination of heat treatment and 10-fold dilution further reduced the inhibition to 4% of the samples. The effects of blood, pH and delay in processing were all evaluated. The inhibition was partly correlated with the pH of the cervical mucosa, but was not shown to be correlated with blood contamination.

In another example, the detection of viral nucleic acids in intraocular fluids and tissues by PCR has become increasingly important in clinical ophthalmology. While much attention has been directed towards minimising false positive reactions resulting from specimen contamination or product carryover, relatively little attention has been given to the causes of false negative PCR results. A PCR inhibitor has been reported to occur in normal aqueous and vitreous intraocular fluids that can produce false negative results.[15] As little as 0.5 μl of vitreous fluid and 20 μl of aqueous fluid can completely inhibit DNA amplification. This inhibition was shown not to be primer specific, nor was it due to chelation of Mg^{2+} ions or DNase activity in the ocular fluid. The inhibitor was completely resistant to boiling for 15 min, but the inhibitory effects were completely removed by a single chloroform–isopentyl alcohol extraction. The extent of PCR inhibition also depended on the type of thermostable DNA polymerase used in the reaction. *Taq* DNA polymerase was very sensitive to the inhibitor, while thermostable DNA polymerases from *Thermus thermophilus* HB-8 (*Tth*) and *Thermus flavus* (*Tfl*) were completely resistant. Thus the inhibitory effects of intraocular fluids on PCR can be removed by diluting the specimen, by chloroform extraction or by using *Tth* or *Tfl* DNA polymerases.

6.2.2.5 Faeces

Monteiro and colleagues developed a model system to study inhibitors present in faeces which prevent the use of PCR for the detection of *Helicobacter pylori*.[16] Inhibitors in faeces are complex polysaccharides, possibly originating from vegetable material in the diet. To detect pathogenic viruses in animal faecal specimens by polymerase chain reaction assays, it is important to remove or inactivate these PCR inhibitory substances. The cationic surfactant Catrimox-14 was successfully used during extraction to detect pathogenic viruses in faecal specimens from a variety of animals.[17] By extraction of viral DNA in the presence of this cationic surfactant, the PCR assay could detect canine parvovirus in all faecal specimens prepared from 13 kinds of animals, i.e. cat, chicken, cow, dog, gerbil, goat, hamster, horse, mouse, pig, rat, rabbit and sheep.

6.2.2.6 Food Samples

Studies have been carried out to investigate interference of the PCR by substances found in clinical material[18] and by detergents.[4] The same consideration needs to be given to the application of PCR to the detection of bacterial pathogens in food. PCR permits high throughput analyses once the purified

DNA is available. Obtaining DNA from the selected tissue or microorganism is often the time-limiting factor to fully exploiting the true potential of the technology. In designing and optimising new generation microbiological assays, the detection is often severely impaired when applied to natural samples. It is therefore important to take into account the fact that any component of the food sample, the growth media or the DNA extraction solutions may influence the effectiveness of the PCR, typically by reducing the detection limits owing to the effect of inhibitory components.

There are various reports in the literature on the interference of PCR by food components, including, for example, a proteinase found in milk which was thought to inhibit PCR by degrading the *Taq* DNA polymerase.[19] Cheese components that reduce the sensitivity of detection of *Listeria monocytogenes* by PCR have also been identified.[20]

Other workers have described a survey of the influence on PCR of a large number of compounds found in food, in media used for selective propagation of food-borne pathogens and in DNA extraction methods.[6] PCR was found to be sensitive to large volumes of complex food samples containing, for example, high amounts of fat and protein. However, the amount of homogenised food sample that could be added directly to the PCR varied considerably with the type of food product. Soft cheeses, for example, completely inhibited the PCR at all concentration levels. The strong inhibitory effect of the cheeses on the PCR may partly be explained by the presence of proteases which can destroy the DNA polymerase structure.

We have also previously reported that dairy products are particularly inhibitory to PCR due to the high levels of calcium ions present.[21] The addition of elevated levels of magnesium to the amplification reaction has been shown to reduce the inhibitory effect of calcium. It appears that there is a competitive interaction between calcium and magnesium ions for a polymerase binding site, with significant effects on the PCR results depending on the concentration of each ion present. Magnesium ions are a vital constituent of PCR, so any interference from a competing divalent cation, such as calcium, decreases the efficiency of the reaction.

In an attempt to overcome the inhibitory effects arising from food matrices, various solutions are beginning to emerge that allow the concentration of target organisms and their subsequent separation from food components. These include the use of lysozyme, proteinase K, detergents, boiling, centrifugation, filtration, DNA affinity purification columns and magnetic beads coated with specific antibodies or lectins.[22] An alternative approach is the direct extraction of bacterial nucleic acids from foods.[23,24] However, since such assays are based on a variety of technologies and formats, their performance in different food matrices can be highly variable.

6.2.2.7   *Culture Media*

In developing diagnostic tests for food pathogens, it has been found that the media used for proliferation often interferes with PCR.[6] The effect of media

Table 6.6 *The effect of bacterial enrichment media on PCR*[6]

Medium	Volume added to PCR		
	10μl	*5 μl*	*1 μl*
Peptone	+	n.d.	n.d.
Brain heart infusion(BHI)	(+)	+	n.d.
Modified enrichment medium (MEB)	(+)	+	n.d.
Fraser	−	+	n.d.
Listeria enrichment broth (LEB)	+	n.d.	n.d.
Phosphate buffered saline (PBS)	−	+	n.d.
Modified Rappaport (MRB)	−	n.d.	+
Rappaport (RB)	n.d.	(+)	+

n.d., not determined; +, no inhibition; (+), some inhibition; − complete inhibition.

commonly used for enrichment of bacteria from food samples, when it is added directly to a PCR mixture with a total volume of 100 μl, is summarised in Table 6.6. In order to analyse if the inhibitory effect of a specific medium could be attributed to specific factors, the effects on PCR of individual medium components were also tested and reported by these workers.

6.2.2.8   Wooden Toothpicks

The PCR can be used for rapid analysis of recombinant DNA directly from bacterial or yeast colonies or viral plaques. In a particularly convenient modification of this procedure, colonies are transferred directly into a complete PCR and cells are lysed during an extended denaturation step prior to cycling. However, it has been reported that the wooden toothpicks used to transfer colonies inhibit PCR reactions containing low amounts of *Taq* DNA polymerase.[25] A comparison of the effects of wooden and plastic toothpicks on the PCR was performed, and the differences observed were attributed to inhibition caused by the wooden toothpicks. The nature of the inhibitor is unknown, but is probably intrinsic to the wood since no chemicals are known to be added during toothpick production.

6.2.2.9   Plant Polysaccharides

Polysaccharides are common contaminants of DNA extracted from plant tissues. Most conventional plant DNA purification methods can remove proteins, but may not be effective for the removal of polysaccharides. Plant polysaccharides may not, however, present as large a problem for PCR as has previously been found for restriction enzyme digests.[26] The effects of neutral and acidic plant polysaccharides on PCR amplification have been investigated.[27] The neutral polysaccharides (arabinogalactan, dextran, gum guar, gum locust bean, inulin, mannan and starch) did not inhibit PCR amplification of spinach DNA. However, of the acidic polysaccharides investigated, two (dextran sulfate and gum ghatti) did inhibit the PCR, whereas the other acidic polysaccharides tested (carrageenan, gum karaya, pectin and xylan) did not cause inhibition.

The addition of 0.5% Tween 20 was shown to reverse the inhibitory effects of gum ghatti (polysaccharide:DNA ratio of 500:1). The inhibitory effect of dextran sulfate (50:1) could be reversed by the addition of Tween 20 (0.25% or 0.5%), DMSO (5%) or poly(ethylene glycol) 400 (5%), but none of these were effective at a 100:1 ratio of dextran sulfate:DNA. The inhibitory nature of some polysaccharides with free acidic groups is further demonstrated by contrasting dextran and dextran sulfate. Dextran (neutral) had no interfering effects at a 500:1 ratio, whereas dextran sulfate was very inhibitory.

Genomic DNA isolated from plants is known to contain higher levels of phenolic compounds and polysaccharides than DNA purified from animal cells. Phenolic compounds are especially troublesome because they oxidise readily during homogenisation, irreversibly interact with proteins and nucleic acids and, consequently, hinder molecular analysis. Inclusion of adjuvants such as DMSO and Tween 20 are reported to counteract the inhibitory effects on PCR by some plant acidic polysaccharides.[27]

6.2.2.10   *Polyphosphates in Fungi*

Genetic, molecular, ecological and evolutionary analyses of filamentous fungi often require the extraction and enzymatic manipulation of nuclear and/or mitochondrial DNA. Many of the techniques commonly used in these studies require that the DNA is amenable to restriction enzyme digestion and/or PCR amplification. Unlike bacterial DNA, extraction of DNA from filamentous fungi is more labour intensive and the quality of the DNA can be compromised by contamination with fungal metabolites.[28] Polyphosphates are ubiquitous in the fungal kingdom and may be responsible for many of the technical difficulties in performing molecular analyses on fungi.

6.2.2.11   *Humic Acid*

While PCR is an attractive method for identifying bacteria, there are a number of problems associated with field samples that must be addressed before successful detection can be achieved. Environmental samples, such as soil, can contain a number of different bacterial species, requiring the isolation of a large amount of sample DNA to enable the detection of individual species. Depending on the source of the sample, various inhibitors of PCR amplification may be present, such as humic and fulvic acids. DNA can also be difficult to isolate in an amplifiable form from field samples owing to the presence of heavy metal ions, or other DNA-damaging agents.

Previous methods for removing PCR inhibitors from soil suspensions have involved the use of Sephadex spin columns.[29] Other methods for DNA purification such as caesium chloride gradient centrifugation and agarose gel electrophoresis are also widely used. The interfering effect of humic acids in PCR-amplified DNA detection can also be eliminated by a calcium chloride precipitation step.[30]

It has been suggested that the inhibitor(s) in soil prevent amplification by binding to the polymerase or to the target DNA. This could potentially be overcome by introducing another component into the reaction with a higher affinity for the inhibitor, thereby preventing their interaction with the polymerase or the target DNA. A series of proteins has been tested for their ability to overcome PCR inhibition by soil suspensions, namely soybean trypsin inhibitor, carbonic anhydrase, glucose 3-phosphate dehydrogenase, ovalbumin, phosphorylase B, β-galactosidase, IgG (bovine), myosin and BSA.[31] Only carbonic anhydrase, ovalbumin, BSA and myosin were found to overcome PCR inhibition by soil suspensions. BSA was found to be the most suitable additive for this purpose.

6.2.2.12 Pollen

After a seasonally recorded complete breakdown of a PCR-based human leucocyte antigen test, a strong and previously unknown PCR inhibitor was identified.[32] This inhibitor was identified as spring pollen. Microscopic evaluation concluded that even 10 or fewer pollen grains per PCR tube may completely inhibit the reaction. Pollen contains a broad variety of biological substances, especially enzymes. Thus enzymatic digestion of a component of the PCR mix may cause the inhibition.

6.2.3 A Review of Approaches for Overcoming Inhibition and Enhancing PCR

Many actions can alleviate or overcome PCR inhibition, some of which have already been mentioned with respect to specific inhibitors (Section 6.2.2.1). A summary of these actions is given in Table 6.7. More specific details of PCR additives and cosolvents are discussed in this section. There is often a balance to be achieved in both reducing inhibition and enhancing the PCR.

The addition of cosolvents has been shown to improve PCR amplification efficiency or specificity in many cases.[33] DMSO, formamide, glycerol and TMAC can improve the specificity and/or efficiency of amplification by influencing template melting, primer annealing properties and *Taq* DNA polymerase activity and thermostability. It should also be noted that all of these cosolvents can be inhibitory at elevated levels; therefore careful optimisation of their use is recommended.

Landre and co-workers[33] remind us that, in some cases, the specificity of PCR may only appear to be improved with the use of cosolvents. If, for example, the enzyme is partially inhibited or other parameters of the PCR adversely affected, less overall amplification of both intended and unintended sequences may

Table 6.7 *Overcoming inhibition and enhancing PCR*

Action to overcome inhibition or enhance PCR	Putative result
DNA clean-up	Removes inhibitors by employing additional purification procedures
Dilution of DNA-containing solution	Reduces concentration of inhibitors added to a PCR reaction. Only appropriate if a sufficient concentration of DNA is available
Heat treatment (5–15 min at 95 or 100 °C)	Inactivates proteases and DNases present in DNA extract
Increase *Taq* DNA polymerase concentration	May overcome inhibition due to enzyme inactivation or by successfully competing with agents that chelate essential enzyme co-factors
Use alternative DNA polymerase enzyme	Alternative DNA polymerases (Section 5.3.5) may be affected differently by inhibitors or may allow the use of higher denaturation temperatures
Add DNA-stabilising cosolvent (TMAC)	Increases T_m by stabilising AT basepairs. May be useful for reducing non-specific products from AT-rich targets by allowing the annealing temperature to be increased[34]
Add DNA-destabilising cosolvent (DMSO, formamide, betaine, glycerol)	Improves the amplification of GC-rich targets by reducing the T_m of DNA
Add non-ionic detergents (Tween 20)	Can reduce secondary structures and stabilise *Taq* DNA polymerase
Add protein agents (BSA, gp32)	Can quench protease activity or preferentially bind inhibitors
Decrease $MgCl_2$ concentration	$MgCl_2$ strongly stabilises the DNA duplex; therefore a reduction may help achieve complete denaturation[6]
Increase $MgCl_2$ concentration	Can increase the amount of free magnesium, and preferentially complete with inhibitory ions for *Taq* DNA polymerase binding site
Add polyamines (spermine)	Stabilises DNA and possibly stabilises enzyme activity
Substitute nucleotide with analogue (c7dGTP)	Reduces secondary structure and non-specific product formation

occur. This may give the impression of reducing the production of non-specific products, whereas in reality a reduction in yield has occurred without any real improvement in reaction specificity. The PCR enhancing abilities of each additive are largely concentration and sample or template dependent. The concentrations quoted can therefore only act as a guideline for use, but may offer a good range for optimisation experiments.

6.2.3.1 Tetramethylammonium Chloride

The use of low annealing temperatures can often lead to mis-priming events, resulting in unacceptably high levels of non-specific DNA amplification. To overcome this problem, the T_m can be increased by the addition of TMAC to the reaction mixture, thus improving the specificity of primer annealing.[34] At 15 mM or more, an increase in the PCR product yield was observed, culminating at 60 mM, while at concentrations greater than 150 mM, TMAC completely inhibited the reaction. Chevet and co-workers concluded that TMAC offers both an enhancement in yield and specificity of the PCR products, while being sequence independent.

There is a limit to the amount of TMAC that can be added to a reaction before the salt content of the buffer becomes inhibitory. However, the amount of TMAC and therefore the enhancing effect can be maximised by replacing all the KCl in the PCR buffer with TMAC.

6.2.3.2 Betaine

Betaine has been reported as a versatile, novel cosolvent for use in PCR.[35] It is non-toxic, inexpensive and has a lower viscosity than glycerol. Betaine binds and stabilises AT base pairs whilst destabilising GC base pairings, resulting in a net specific destabilisation of GC-rich regions.[36] The use of dsDNA destabilising reagents, such as betaine, has been widely recommended to improve the PCR of GC-rich templates. The effect of various cosolvents (DMSO, glycerol, trehalose, TMAC and betaine) on competitive PCR has been investigated at LGC.[37] Whilst the majority of cosolvents investigated had no visible effect on reducing the preferential amplification of the target, the addition of betaine successfully allowed the GC-rich mimic to be competitively amplified.

Unlike TMAC, betaine is not a salt and can therefore be added to PCR at greater concentrations to allow maximal improvement of amplification. Another advantage of betaine lies in its ability to increase PCR tolerance to heparin contamination. Heparin is a common contaminant in DNA preparations from clinical blood samples and certain cell types, as previously mentioned (Section 6.2.2.1). It has been shown that the addition of betaine improved the tolerance of PCR to heparin by 2.5-fold to 25-fold.[35]

6.2.3.3 Glycerol

Glycerol can protect the DNA polymerase enzyme when included in a PCR assay and reduce T_m by destabilising DNA.[33] Possibly because of these properties, at 10% it has proved helpful in amplifying long (up to 2500 bp) PCR products.[38]

6.2.3.4 DMSO

DMSO has been shown to improve the PCR amplification of DNA with a complex secondary structure[39] and may be useful for the amplification of GC-rich targets. When adding DMSO at 10%, the concentration of *Taq* DNA polymerase may need to be increased to compensate for nearly 50% enzyme inhibition.[40] The addition of 5% DMSO was found to reduce the inhibitory effect of some acidic plant polysaccharides.[27]

6.2.3.5   Formamide

It has been observed that *Taq* DNA polymerase is stable in low concentrations of formamide (10%), provided glycerol is present (10%). The effect of this combination is to decrease the T_m of the DNA duplex without harming the enzyme.[5] Landre and co-workers[33] also reported that the use of both glycerol (1%) and formamide (1%) together and formamide alone (1 and 2.5%) in a PCR facilitated both amplification at lower denaturation temperatures and amplification from templates with a high GC content.

6.2.3.6   Non-ionic Detergents

Non-ionic detergents are frequently added to buffers to help stabilise enzymes and therefore may assist in a PCR for this reason. Tween 20 (0.5%) reversed the inhibitory effects of some acidic plant polysaccharides.[27]

6.2.3.7   Protein Additives: Bovine Serum Albumin and gp32

Protein based additives can be added to a PCR, either to act as a substrate for protease activity[19] or to bind inhibitors. Perhaps the two most common proteins used for this purpose are BSA and gp32; the latter is a thermostable, single-stranded binding protein which Tijssen[5] recommends using at 3 μg per 100 μl reaction. The benefits of adding either BSA or gp32 to PCRs containing typical inhibitors have been evaluated. Kreader[41] found that a 10–1000-fold increase of the inhibitors $FeCl_2$, hemin, fulvic acids, humic acid, tannin acid or extracts from faeces, freshwater or marine water could be accommodated in the PCR when either 400 ng/μl of BSA or 150 ng/μl of gp32 were included in the reaction. The study also revealed that neither BSA nor gp32 significantly reversed the effects of minimum inhibitory levels of bile salts, bilirubin, EDTA, NaCl, SDS or Triton X-100.

   In separate studies, the incorporation of BSA (at approximately 1 mg/ml) prevented the inhibition of PCR by the heme compounds[11] and, present at an optimum concentration of 67 mM, was necessary for the successful amplification of some modern and the majority of ancient museum samples.[42]

6.2.3.8   Spermine/Spermidine

Spermidine [*N*-(3-aminopropyl)-1,4-butanediamine] is a polyamine that is routinely included in restriction enzyme digestions to improve the cleavage of DNA. A review of the use of spermidine[43] reports that it counteracts the inhibitory effects of contaminants co-isolated with DNA and consequently permits complete digestion of the DNA at lower enzyme concentrations. Experiments *in vitro* have shown that spermidine has a high affinity for nucleic acids and neutralises at least part of the negative charges in the phosphate backbone, thereby stabilising DNA and RNA. Polyamines are also known to stimulate the activity of enzymes involved in nucleic acid metabolism, such as DNA and RNA polymerases and topoisomerases.[43]

Effects on PCR of the polyamines spermine and spermidine, employed either singly or together, have also been reported with optimum enhancing concentrations ranging from 0.4 to 0.6 mM.[44] When tested separately, spermidine was more efficient than spermine in promoting amplification.

In addition, spermidine has been shown significantly to enhance PCR amplification of plant DNA where DMSO, formamide and Tween 20 failed to produce the expected PCR product.[43] Wan and Wilkins report that 0.2–1 mM spermidine exerted a significant enhancing effect on PCR amplification of DNA, producing higher specificity and reproducibility, yet allowing simplified DNA isolation procedures to be employed.[43]

However, the use of a DNA isolation method containing polyamines in the extraction buffer involves the risk of carrying over the polyamines to the isolated DNA at a concentration that could affect the template properties during the PCR. It is possible that the binding of excessive polyamines affects the apparent availability of the DNA template. Hence the concentration of polyamines, such as spermine and spermidine, may be critical in either enhancing or inhibiting PCR amplification.

6.2.3.9 Nucleotide Analogues

Incorporation of the nucleotide analogue 7-deaza-2'-deoxyguanosine triphosphate (c7dGTP), in addition to deoxyguanosine triphosphate (dGTP), at a ratio of 3:1 has been found to help destabilise secondary structures of DNA, improve the amplification of GC-rich targets and reduce the formation of non-specific products.[45]

6.3 Concluding Comments

Much has been written describing the effects of PCR inhibitors and enhancers on the amplification process, with almost as many outcomes as there are applications of the technique.

- An ever increasing number of inhibitors and enhancers have been found to affect the PCR amplification of DNA, originating from a wide range of biological specimens.
- Many crude preparations contain inhibitory substances, although the exact mechanisms of inhibition are not always known.
- There is a very marked variation in the way in which chemicals and reagents affect PCR amplification, often making difficult the successful transfer of conditions from one assay to another.

Different samples are likely to possess varying concentrations of a wide range of inhibitory components and it is unlikely that any single enhancer will guarantee the reversal of such inhibition or result in successful amplification in all cases. Each situation in which a PCR is to be adopted must be rigorously optimised and controlled to provide confidence in the results obtained.

6.4   References

1. Saiki, R. K., Gelfand, D. H., Stoffel, S., Scharf, S. J., Higuchi, R., Horn, G. T., Mullis, K. B. and Erlich, H. A. 1988. Primer-directed enzymatic amplification of DNA with a thermostable DNA polymerase. *Science* **239**: 487–491.
2. Cone, R. W., Hobson, A. C. and Huang, M. W. 1992. Coamplified positive control detects inhibition of polymerase chain reactions. *J. Clin. Microbiol.* **30**: 3185–3189.
3. Gelfand, D. H. 1989. *Taq* DNA polymerase. In: PCR Technology, Principles and Applications for DNA Amplification (ed. Erlich, H. A.), pp. 17–22. Stockton Press, New York.
4. Weyant, R. S., Edmunds, P. and Swaminathan, B. 1990. Effects of ionic and nonionic detergents on the *Taq* polymerase. *BioTechniques* **9**: 308–309.
5. Tijssen, P. 1993. Hybridization with nucleic acid probes. In: Laboratory Techniques in Biochemistry and Molecular Biology (ed. van der Vliet, G. M.), vol. 24, pp. 191–195. Elsevier.
6. Rossen, L., Norskov, P., Holmstrom, K. and Rasmussen, O. F. 1992. Inhibition of PCR by components of food samples, microbial diagnostic assays and DNA-extraction solutions. *Int. J. Food Microbiol.* **17**: 37–45.
7. Hoppe, B. L., Conti-Tronconi, B. M. and Horton, R. M. 1992. Gel-loading dyes compatible with PCR. *BioTechniques* **12**: 679–680.
8. Gilgen, M., Hofelein, C., Luthy, J. and Hubner, P. 1995. Hydroxyquinoline overcomes PCR inhibition by UV-damaged mineral oil. *Nucleic Acids Res.* **23**: 4001–4002.
9. Dohner, D. E., Dehner, M. S. and Gelb, L. D. 1995. Inhibition of PCR by mineral oil exposed to UV irradiation for prolonged periods. *BioTechniques* **18**: 964–966.
10. Taylor, A. C. 1997. Titration of heparinase for removal of the PCR inhibitory effect of heparin in DNA samples. *Mol. Ecol.* **6**: 383–385.
11. Akane, A., Matsubara, K., Nakamura, H., Takahashi, S. and Kimura, K. 1994. Identification of the heme compound copurified with deoxyribonucleic acid (DNA) from bloodstains, a major inhibitor of polymerase chain reaction (PCR) amplification. *J. Forensic Sci.* **39**: 362–372.
12. Comey, C. T., Koons, B. W., Presly, K. W., Smerick, J. B., Sobieralski, C. A., Stanley, D. M. and Baechtel, F. S. 1994. DNA extraction strategies for amplified fragment length polymorphism analysis. *J. Forensic Sci.* **39**: 1254–1269.
13. Akane, A. 1996. Hydrogen peroxide decomposes the heme compound in forensic specimens and improves the efficiency of PCR. *BioTechniques* **21**: 392–394.
14. Verkooyen, R. P., Luijendijk, A., Huisman, W. M., Goessens, W. H., Kluytmans, J. A., van Rijsoort-Vos, J. H. and Verbrugh, H. A. 1996. Detection of PCR inhibitors in cervical specimens by using the Amplicor *Chlamydia trachomatis* assay. *J. Clin. Microbiol.* **34**: 3072–3074.
15. Wiedbrauk, D. L., Werner, J. C. and Drevon, A. M. 1995. Inhibition of PCR by aqueous and vitreous fluids. *J. Clin. Microbiol.* **33**: 2643–2646.
16. Monteiro, L., Bonnemaison, D., Vekris, A., Petry, K. G., Bonnet, J., Vidal, R., Cabrita, J. and Megraud, F. 1997. Complex polysaccharides as PCR inhibitors in faeces: *Helicobacter pylori* model. *J. Clin. Microbiol.* **35**: 995–998.
17. Uwatoko, K., Sunairi, M., Yamamoto, A., Nakajima, M. and Yamaura, K. 1996.

Rapid and efficient method to eliminate substances inhibitory to the polymerase chain reaction from animal fecal substances. *Vet. Microbiol.* **52**: 73–79.

18. Panaccio, M. and Lew, A. 1991. PCR based diagnosis in the presence of 8% (v/v) blood. *Nucleic Acids Res.* **19**: 1151.

19. Powell, H. A., Gooding, C. M., Garrett, S. D., Lund, B. M. and McKee, R. A. 1994. Proteinase inhibition of the detection of *Listeria monocytogenes* in milk using the polymerase chain reaction. *Lett. Appl. Microbiol.* **18**: 59–61.

20. Herman, L. and De Ridder, H. 1993. Cheese components reduce the sensitivity of detection of *Listeria monocytogenes* by the polymerase chain reaction. *Neth. Milk Dairy J.* **47**: 23–29.

21. Bickley, J., Short, J. K., McDowell, D .G. and Parkes, H. C. 1996. Polymerase chain reaction (PCR) detection of *Listeria monocytogenes* in diluted milk and reversal of PCR inhibition caused by calcium ions. *Lett. Appl. Microbiol.* **22**: 153–158.

22. Kroll, R. G. 1993. Microbiological analysis of foods. In: DNA Probes (eds. Keller, G. H. and Manak, M. M.), pp. 565–588. Stockton Press, New York.

23. Dickinson, J., Kroll, R. G. and Grant, K.A. 1995. The direct application of the polymerase chain reaction to DNA extracted from foods. *Lett. Appl. Microbiol.* **20**: 212–216.

24. Makino, S., Okada, Y. and Maruyama, T. 1995. A new method for direct detection of *Listeria monocytogenes* from foods by PCR. *Appl. Environ. Microbiol.* **61**: 3745–3747.

25. Lee, A. B. and Cooper, T. A. 1995. Improved direct PCR screen for bacterial colonies: wooden toothpicks inhibit PCR amplification. *BioTechniques* **18**: 225–226.

26. Do, N. and Adams, R. P. 1991. A simple technique for removing plant polysaccharide contaminants from DNA. *BioTechniques* **10**: 162–166.

27. Demeke, T. and Adams, R. P. 1992. The effects of plant polysaccharides and buffer additives on PCR. *BioTechniques* **12**: 332–334.

28. Rodriguez, R. J. 1993. Polyphosphate present in DNA preparations from filamentous fungal species of *Colletotrichum* inhibits restriction endonucleases and other enzymes. *Anal. Biochem.* **209**: 291–297.

29. Tsai, Y. L. and Olsen, B. H. 1992. Rapid methods for separation of bacterial DNA from humic substances in sediments for polymerase chain reaction. *Appl. Environ. Microbiol.* **58**: 2292–2295.

30. Ernst, D., Kiefer, E., Drouet, A. and Sandermann, H. 1996. A simple method of DNA extraction from soil for detection of composite transgenic plants by PCR. *Plant Mol. Biol. Reporter* **14**: 143–148.

31. McGregor, D. P., Forster, S., Steven, J., Adair, J., Leary, S. E. C., Leslie, D. L., Harris, W. J. and Titball, R. W. 1996. Simultaneous detection of microorganisms in soil suspension based on PCR amplification of bacterial 16S rRNA fragments. *BioTechniques* **21**: 463–471.

32. St Pierre, B., Neustock, P., Schramm, U., Wilhelm, D., Kirchner, H. and Bein, G. 1994. Seasonal breakdown of polymerase chain reaction. *Lancet* **343**: 673.

33. Landre, P. A., Gelfand, D. H. and Watson, R. M. 1995. The use of cosolvents to enhance amplification by the polymerase chain reaction. In: PCR Strategies (eds. Innis, M. A., Gelfand, D. H. and Sninsky, J. J.), pp. 3–16. Academic Press, San Diego, CA.

34. Chevet, E., Lemaitre, G. and Katinka, M. D. 1995. Low concentrations of tetramethylammonium chloride increase yield and specificity of PCR. *Nucleic Acids Res.* **23**: 3343–3344.

35. Weissensteiner, T. and Lanchbury, J. S. 1996. Strategy for controlling preferential amplification and avoiding false negatives in PCR typing. *BioTechniques* **21**: 1102–1108.

36. Rees, W. E., Yager, T. D., Korte, J. and von Hippel, P. 1993. Betaine can eliminate the base pair composition dependence of DNA melting. *Biochemistry* **32**: 137–144.

37. McDowell, D. G., Burns, N. A. and Parkes, H. C. 1998. Localised sequence regions

possessing high melting temperatures prevent the amplification of a DNA mimic in competitive PCR. *Nucleic Acids Res.* **26**: 3340–3347.

38. Smith, K. T., Long, C. M., Bowman, B. and Manos, M. M. 1990. Using cosolvents to enhance PCR amplification. *Amplifications* (Perkin Elmer Cetus) **5**: 16–17.

39. Shen, W. H. and Hohn, B. 1992. DMSO improves PCR amplification of DNA with complex secondary structure. *Trends Genetics* **8**: 227.

40. Guevara-Garcia, L., Herrera-Estrella, L. and Olmedo-Alvarez, G. 1997 Cloning from genomic DNA and production of libraries. In: Plant Molecular Biology — A Laboratory Manual (ed. Clark, M. S.), p. 62. Springer, Berlin.

41. Kreader, C. A. 1996. Relief of amplification inhibition in PCR with bovine serum albumin or T4 gene 32 protein. *Appl. Environ. Microbiol.* **62**: 1102–1106.

42. Cooper, A. 1994. DNA from museum specimens. In: Ancient DNA (eds. Herman, B. and Hummel, S.). Springer, New York.

43. Wan, C. and Wilkins, T. A. 1993. Spermidine facilitates PCR amplification of target DNA. *PCR Methods Applications* **3**: 208–210.

44. Ahokas, H. and Erkkila, M. J. 1993. Interference of PCR amplification by the polyamines, spermine and spermidine. *PCR Methods Applications* **3**: 65–68.

45. McConlogue, L., Brow, M. D. and Innis, M.A. 1988. Structure-independent DNA amplification by PCR using 7-deaza-2'-deoxyguanosine. *Nucleic Acids Res.* **16**: 9869.

CHAPTER 7

Random Amplified Polymorphic DNA (RAPD) Analysis

GINNY C. SAUNDERS AND DANIEL HOPKINS

7.1 Introduction

Random amplified polymorphic DNA (RAPD) profiling is an application of the standard PCR used to detect DNA polymorphic differences that exist between individuals and species.[1,2] A single 10 bp primer (10mer) of arbitrary sequence is used to amplify DNA fragments from a genomic template. This simple technique has been widely used for analysis as it has a significant advantage over other profiling techniques, namely that no prior knowledge of the genome sequence is required owing to the arbitrary nature of the technique.

Each PCR product is of an unknown sequence and is amplified when the primer anneals to the genomic DNA fulfilling three criteria (see Figure 7.1). Firstly, two primers must anneal on opposite strands; secondly, they must be in

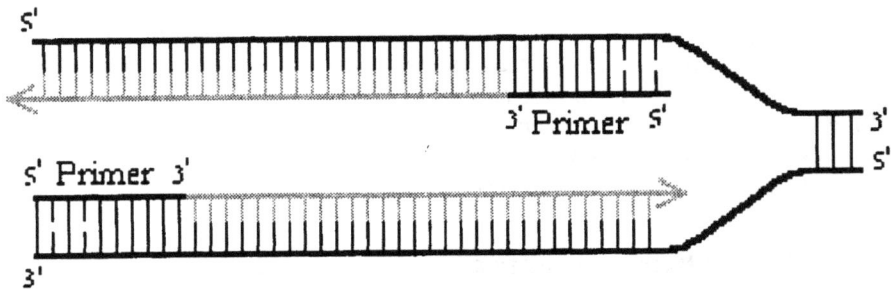

Figure 7.1 *Schematic diagram to illustrate the three criteria required for the annealing of a primer to facilitate the amplification of DNA to produce a RAPD profile. The primer must (i) anneal on opposite strands, (ii) in opposing orientations (3' ends internal) and (iii) within a range of less than approximately 3 kbp. Extension from the primer is shown in grey. Mis-matching of the primer to the original template tends to be more tolerated towards the 5' end of the 10mer*

opposing orientations (with the 3′ ends flanking the intervening sequence); and thirdly, they must anneal within a given distance (no greater than approximately 3 kbps). The last criterion is to ensure that the *Taq* polymerase can complete the synthesis of the intervening sequence within the time allowed for primer extension and that the length of the second strand synthesis is within the capacity of the polymerase.

Synthesis of the complementary strand occurs due *Taq* DNA polymerase-mediated extension of the primer from the 3′ end by the addition of dNTPs. After the first few cycles of PCR, these newly synthesised segments will become predominant over the original genomic template, and in theory, will be amplified exponentially. The event described in Figure 7.1 can occur at any number of sites in a genome, from nil upwards. The resulting amplified products, when electrophoresed on an agarose or acrylamide gel stained with ethidium bromide or silver, produce the characteristic RAPD profile.

RAPD analysis is one of the three core techniques that employ an arbitrarily primed amplification approach, the others being arbitrarily primed PCR (AP-PCR) and DNA amplification fingerprinting (DAF) (see Table 7.3).

In reality, the primer may anneal to sequences on the template that are not 100% homologous. It is thought that mis-matching of up to 30% (i.e. 3 in 10 bases) is tolerated in the first few cycles of amplification, with the greatest homology being required at the 3′ end of the primer[3] in order to allow the *Taq* polymerase to bind and commence primer extension. Mis-priming is made possible and stabilised by the low temperature of primer annealing.[4] Primer mis-matching events are required when working with species of relatively small genome size such as fungi, since, statistically, the number of bands resulting from an exact match of a random 10mer annealing to a yeast genome of 1.6×10^4 kbp is only 0.029.[5] Of course, after preliminary cycles, when the PCR product containing the exact primer sequence out-numbers the original template, the primers will anneal with 100% homology.

Polymorphic amplified products are seen as bands of differing sizes on a gel after electrophoretic separation. When comparing isolates, an isolate is either positive or negative with respect to the presence or absence of a given band. Polymorphic bands are generated as a result of a variation in the DNA sequence of the intervening region between priming sites, or a variation in, or lack of, the primer site sequence. These variations may be in the form of base substitutions, insertions or deletions varying in length from 1 bp to many hundreds of bases.

7.1.1 RAPD Reproducibility and Repeatability

The random nature and low specificity of the RAPD reaction can leave this methodology prone to reduced reliability and reproducibility, resulting in variation of analytical data. Owing to the variation of RAPD results obtained when using different thermal cyclers and when attempting to transfer protocols between differing pieces of equipment (discussed in Chapter 5, Figures 5.2 and 5.3), obtaining an acceptable level of reproducibility between laboratories may remain elusive. In spite of this obvious limitation to the technology and the further validation issues discussed in this chapter, the RAPD methodology remains popular. A literature search of common databases (BIOSIS preview, CAB abstracts, Elsevier biobase, Pascal, Medline, Scisearch, Agricola and CA search), covering the 14 months prior to February 1998, identified approximately 500 unique articles with RAPD in the title. The authors therefore feel justified in including a chapter on RAPD methodology in a publication on quality, as we believe that much can still be achieved by way of improving the repeatability of the data obtained and raising awareness concerning the limitations of RAPD analysis.

Related technologies are presented in Section 7.5 which may offer viable alternatives to users of RAPD analysis and many of the issues covered in this Chapter also apply to these technologies.

This chapter briefly describes a 'typical' RAPD reaction and then aims to highlight critical points in the methodology which, when addressed, may improve the repeatability of profiles produced.

7.2 Applications of RAPD Analysis

The applications of RAPD analysis are vast and the number and variety of uses are still increasing. Information gained from RAPD profiles can be used to determine phylogenetic relationships and undertake population and epidemiological studies of higher and lower eukaryotic species and prokaryotic isolates.

Particular interest has been shown in the investigation of a diverse range of microorganisms, as no previous knowledge of the organism is required and RAPD analysis can differentiate close relatives at the sub-species or even individual level. RAPD profiling has been used for diagnostic and epidemiological studies in human pathogenic species of fungi. This technique was successfully used to distinguish isolates of *Aspergillus fumigatus*, the cause of farmer's lung.[6] RAPD profiling was also used to estimate the age and size of one of the largest organisms on the earth. RAPD patterns of the fungus *Armillaria bulbosa* sampled from a large site in Michigan were found to be invariant, indicating that all the isolates represented a single clone. The age of the clone was estimated to be 1500 years old.[7]

Many species of pathogenic fungi and their plant hosts have also been characterised and identified using RAPD analysis. This form of analysis has provided information on population genetics, epidemiology and pathotype identification.[8]

7.3 RAPD Methodology

There is no single RAPD reaction that can be presented in this text that can be guaranteed to work on all samples, with all primers and using all thermal cyclers. It is therefore advisable to optimise a reaction in-house. Optimisation of PCR components and thermal cycling conditions will depend upon genome size, G/C content of both the primer and template, and template quality.

For a structured approach to RAPD optimisation, Cobb[9] describes the use of Taguchi methods employing orthogonal arrays to combine the various components and concentrations, thereby reducing the number of experiments to be carried out and statistically arriving at optimal conditions.

7.3.1 Extraction of DNA

High quality DNA template is vital to the success of a RAPD reaction, and time invested in optimising a suitable extraction procedure that provides DNA free from contaminants and degradation is time well spent (Chapter 3).

7.3.2 Estimation of DNA Concentration

In order to maintain consistency between samples, the amount of starting template DNA must be determined and optimised. DNA concentration is usually estimated by one of two methods: (i) dilution of the extracted DNA with sterile distilled water to obtain a spectrophotometric reading at 260 nm in the range of $A = 0.1$–1.0 (one A_{260} unit of double stranded DNA is equal to approximately 50 μg/ml) and (ii) electrophoresis of 10, 5 and 1 μl of the DNA on a 2% agarose gel against three samples of known concentration. For more information on determining DNA concentration, see Chapter 4.

7.3.3 Optimisation of PCR-RAPD Cycling Conditions

Tables 7.1 and 7.2 present the RAPD reaction components and conditions used at LGC to produce profiles from fungal isolates, which may be useful as a starting reaction for optimisation. Both tables contain a column that shows the ranges of the various components and conditions that could be explored.

Table 7.1 *The standard PCR reaction lists components employed in our laboratory for the RAPD profiling of fungal species*

Standard RAPD reaction			
1 × RAPD reaction	*μl*	*Final conc.*	*Optimisation ranges — final conc.*
10 × amplification buffer (as recommended by the *Taq* polymerase supplier)	2.5	1 ×	–
$MgCl_2$ (25 mM)	2.5	(2.5 μM)	1.0–6.0 μM
dNTP mixture (1.25 mM)	2.0	(100 μM)	50–250 μM
Taq polymerase (5 units/μl)	0.12	(0.024 U/μl)	0.02–0.1 U/μl
Primer (25 pmol/μl)	1.0	(1 μM)	1–6 μM
Sterile distilled water	11.88		–
DNA (25–100 ng)	5.0	(1–4 ng/μl)	0.2–20 ng/μl
Total volume	25.0		

Table 7.2 *Temperature profile for a RAPD reaction carried out on a Perkin Elmer 2400 or 9600 and possible optimisation ranges for other thermal cyclers*

Standard RAPD reaction		Optimisation range
94 °C for 4 min	1 cycle	94–95 °C for 3–10 min
94 °C for 1 min		92–94 °C for 5–90 s
37 °C for 1 min	35 cycles	32–42 °C for 20 s–2 min
72 °C for 2 min*		72 °C for 30 s–2 min†
72 °C for 7 min	1 cycle	72 °C for 2–10 min
4 °C	hold ∞	

*Ramp rate from 37 to 72 °C set at 30%.
†Ramp rate from 37 to 72 °C set at 10–100%. 25–40 cycles can be used.

RAPD amplification is usually carried out in a 25 μl reaction volume for optimisation purposes.

1. Prepare a master mix of all common PCR reagents (except the DNA) for the number of reactions to be carried out, including a negative control with no DNA, a positive control with characterised template, plus one (to allow for small losses incurred when aliquoting out). Master mixes should be used whenever possible as they will minimise any errors that may be introduced by pipetting small volumes. Add the *Taq* polymerase to the master mix last, briefly vortex the mix and spin down for 10 s.
2. Aliquot the correct volume of the master mix (20 μl in this case) into the appropriate size Eppendorf for the thermal cycler.
3. Add the quantitated, diluted DNA. The DNA solution should be no more than one fifth of the reaction volume.
4. Overlay with mineral oil (25 μl) if the thermal cycler requires it.
5. Briefly spin down the contents of the Eppendorf, before placing in the thermal cycler.
6. Following amplification, briefly spin down the reaction before electrophoresing 5–10 μl of RAPD PCR product plus loading buffer on a 2% agarose gel containing ethidium bromide.

7.3.4 Trouble Shooting During Optimisation

If an insufficient number of RAPD fragments are obtained from a template, or a limited number of primers are available, two primers can be used in combination in a single PCR reaction. Care should be taken to ensure that the two primers do not complement each other in sequence. Profiles obtained from pairwise primers should not be used in addition to profiles obtained from one of the primers individually as there is an increase in the possibility of the same amplification products being produced, therefore reducing the diversity of the profiles.

If no RAPD amplification is obtained, several options are available:

- Ensure that the DNA is completely resuspended following the extraction process, accurately quantitated and free from contaminants.
- If the template DNA is considerably degraded (seen as a smear on an agarose gel), then an alternative extraction method may be required. If the specimen itself is old, degraded or highly processed, then alternative profiling techniques should be explored.
- Adjust the ramp rate between annealing and extension. On the majority of thermal cyclers the default is usually set for the fastest transition time possible. An extended ramp time of a 3 min transition period from 37 to 72 °C may be beneficial, especially when amplifying small genomes which may be dependent on mis-priming events for products to occur.
- The annealing temperature can be lowered down to 32 °C, but often just one or two degrees may make a difference.

If smearing or many shadowy or intense bands are obtained from an amplification, specificity can be improved by:

- Increasing the annealing temperature in steps of 1–2 °C.
- Reducing the $MgCl_2$ concentration to 1.5 mM.
- Reducing the amount of DNA template.
- Reducing the amount of *Taq* polymerase

7.3.5 Visualisation of RAPD Profiles

Profiles found to be most informative in agarose gel analysis are generally those that contain up to 15 fragments, each clearly distinguishable as a clear band on a gel. A profile containing more bands can be difficult to analyse owing to the poor resolving power of the gel matrix and electrophoresis technology. Profiles containing a greater number of bands are more suitably analysed using a polyacrylamide gel matrix, allowing better resolution of co-migrating bands of a similar size.

Several developments in product separation have been successfully applied to RAPD profile analysis. RAPD products produced with either fluorescently labelled nucleotides or primers can be resolved, visualised and accurately sized by employing the ABI 373 or 377 DNA Sequencer, internal size markers and dedicated software.[10]

Improved resolution is achieved using denaturing gradient gel electrophoresis (DGGE) and temperature sweep gel electrophoresis (TSGE), techniques which resolve DNA fragments with respect to both the size and melting properties of the fragment; therefore sequence variation as well as fragment length can be detected.[11,12]

DNA Fragments unique to a RAPD profile can be isolated from the gel matrix and used as a source of species- or isolate-specific probes that can be used for the detection of the organism by hybridisation.[13]

7.3.6 Analysis of RAPD Data

The similarity of profiles can be simply assessed by comparing the number of shared and unique bands between profiles. This can be translated into the similarity coefficient (F) of Nei and Li.[14] A similarity coefficient can have a value between 0 and 1, where 0 indicates that there are no bands in common (low genetic similarity) and 1 indicates identical band patterns (high genetic similarity). An example is given Figure 7.2.

Similarity coefficients between samples are usually displayed in a matrix format, as shown in Figure 7.2. The DNA from samples or the actual samples of a given population can be pooled in order for comparisons to be made between populations.

Other statistical methods to evaluate fragment patterns are available for the

$$F = \frac{2 \text{ (No. of common fragments)}}{\text{(No. of fragments in A)} + \text{(No. of fragments in B)}}$$

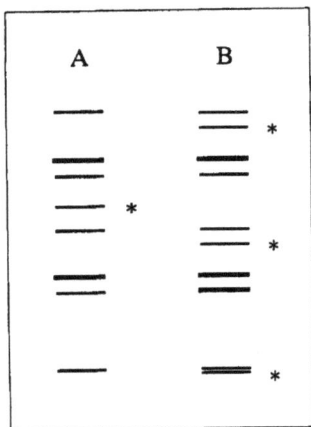

$$F = \frac{2 \times 7}{8 + 10} = 0.78$$

	A	B	C	D
A	-			
B	0.78	-		
C	0.86	0.42	-	
D	0.68	0.38	0.97	-

Figure 7.2 *Example of the method used for determining genetic similarity between samples employing the similarity coefficient of Nei and Li.[14] (* denotes unique fragments)*

assessment of phylogenetic analysis for determining genetic diversity and relatedness, and linkage analysis for gene mapping. These are clearly presented and discussed in Weising *et al.*[8]

When assessing samples for comparison, a decision has to be made whether differences in band intensity should be taken into account. In the sample given in Figure 7.2 the differences are ignored. Reproducibly different intensities may reflect the number of initial target sequences (i.e. single or multi copy), or may result from the co-amplification of multiple heterogeneous fragments of the same size. Alternatively, differences in band intensities can be scored by assuming that one shared fragment exists in both samples, but a second fragment of the same size is present in the more intense sample. Non-reproducible differences in intensity should not be scored in such a way. Should these consistently appear, the quality of the DNA should be investigated.

7.4   Validation Issues of RAPD Analysis

The reproducibility of RAPD profiles has been a controversial subject of discussion almost since the publication of the first descriptions of the RAPD technique and the initial impetus to apply the technique across various sectors. Non-reproducible techniques have obvious limitations, such as a lack of comparability of results between different thermal cyclers and laboratories and, a difficulty in developing a useful database for identification purposes; therefore the validity of the data they produce remains questionable. However, a literature search of RAPD analysis leaves you in little doubt that this technique is still proving a popular tool in analytical and research laboratories. This is, in part, due to the open debate in numerous scientific fora that has improved the understanding of the various causes of non-reproducibility of RAPD profiles and a great effort by a number of workers to promote improvement in these areas. The critical points in RAPD profiling are discussed below.

7.4.1   DNA Quality and Quantity

Poor quality DNA can affect the validity of RAPD data by promoting the non-reproducibility of profiles. This can be due to the presence of endogenous or exogenous inhibitors in the extract, which can affect the quantification of the sample and inhibit the *Taq* polymerase. A high degree of degradation of the template DNA could also inhibit the amplification of some of the larger fragments in the reaction owing to a decrease in the availability of intact target template. Extraction processes that commonly produce degraded DNA (such as 'Chelex' extractions) should therefore be avoided. DNA degradation can be assessed by viewing the DNA on an agarose gel, since degraded DNA will be seen as a smear or will run as low molecular weight nucleic acids. The extracted DNA should ideally appear as high molecular weight bands

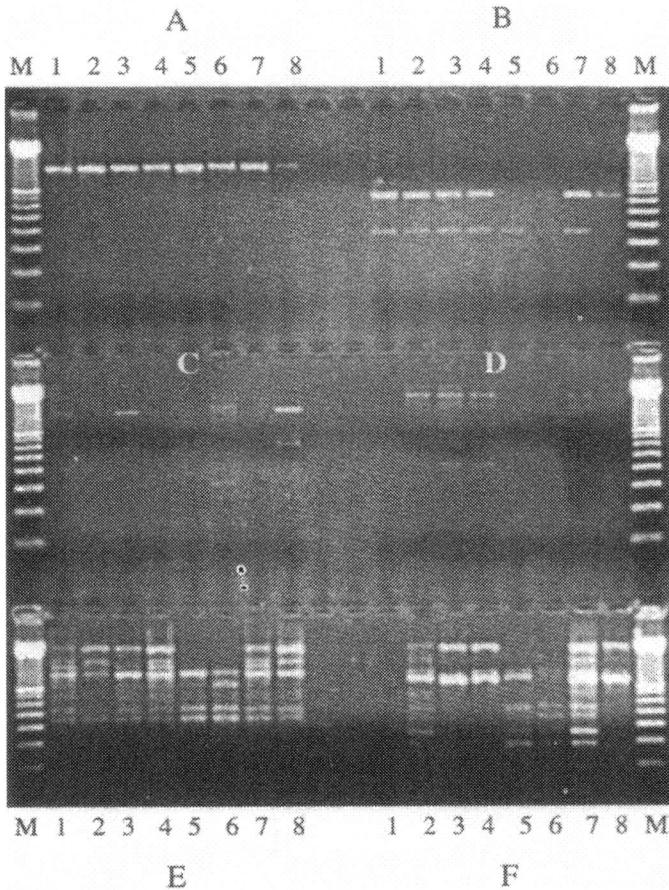

Figure 7.3 *The effect of DNA quality on the reproducibility of RAPD profiles. RAPD profiles were generated using the same conditions. Lanes 1 to 8 are generated from DNA extracted by eight different methodologies. A to F are six different isolates of fungi:* P. chrysoginum *(A),* A. alliacius *(B),* A. niger *(C),* A. foetidus *(D),* S. cerevisiae *140023 (E), and* S. cerevisiae *(F). M represents a 100 bp molecular weight marker*

(running approximately level with the 23 kb marker) when electrophoresed on an agarose gel.

The effect of DNA quality on the reproducibility of the RAPD profile is clearly demonstrated in Figure 7.3. The differences in the RAPD profiles obtained from DNA extracted using different methodologies could be due to differences in the integrity of the DNA or the presence of co-extracted inhibitory factors. Venugopal et al.[4] also noted that RAPD patterns are affected by non-DNA impurities or contaminants. Yu and Pauls[15] have demonstrated that simply centrifuging the DNA before adding it to the PCR reagents can increase reproducibility of profiles by pelleting particulate material.

Dilution of DNA from a typical extraction is usually required to adjust the concentration before addition to the PCR mix. However, if the DNA is present at very low concentrations and has to be used undiluted, as a guideline the volume of DNA should be no more than a fifth of the total reaction volume. This can help minimise the addition of any residual inhibitory factors. The presence of inhibitors in the DNA extract can be identified by performing RAPD reactions using serial dilutions of DNA. Confirmation of the presence of inhibitors in the DNA extract will be established by the emergence of a profile as the extract is diluted.

To ensure that the concentration of template DNA employed in the RAPD reaction is not borderline to acceptability, DNA amounts of at least a two-fold dilution and a two-times concentration should be examined during validation. The profiles obtained should all be comparable and reproducible.

7.4.2   Biological Contamination

All techniques based on PCR should take adequate care to avoid cross contamination of samples and maintain the highest purity of all PCR components. This is particularly true for arbitrarily primed PCRs as potentially all DNA contaminants can generate amplification products with the random primers. Negative controls, containing all the reaction components except the DNA template, are therefore an essential feature of all RAPD experiments.

Occasionally, non-specific bands have been seen to occur in sterile distilled water controls after 35 cycles. A *Taq* DNA polymerase manufacturer has suggested that this may be due to residual host or vector DNA being present in the recombinant *Taq* DNA polymerase preparation. For this reason, we carry out RAPD amplification at a maximum of 35 cycles.

7.4.3   *Taq* Polymerase Enzyme

Stoffel DNA polymerase has been frequently employed for RAPD amplification as it has a high thermal stability. This is perhaps more relevant when RAPD reactions required up to 7 h to complete. With the advent of the latest thermal cyclers, amplification can now be completed in a much shorter time period and standard *Taq* DNA polymerases are sufficient. It should be noted, however, that different profiles can be obtained from the use of different enzymes.[16] Therefore once a RAPD assay has been validated, the enzyme should not be changed without further validation. Meunier and Grimont[17] found that with the use of consistent DNA sample and target concentration, *Taq* DNA polymerase and thermocycler, reproducibility was excellent. However, when different *Taq* DNA polymerases were introduced, reproducibility was not maintained.

7.4.4 Thermal Cyclers

Several groups have reported problems associated with the reproducibility of RAPD profiles. MacPhearson *et al.*[18] found reproducibility a concern when comparing results from three commonly used thermal cyclers, and Penner *et al.*[19] found that a different range of band sizes were amplified when the same template was tested in six different laboratories.

Section 5.3.3 discusses inefficient block heat transfer within a thermal cycler, with the consequence that samples in some positions amplify products less efficiently and therefore less reproducibly. Thermal block inefficiencies should be investigated, particularly on older style blocks and ambient cyclers, especially since the affect on RAPDs is often much more pronounced than a simple specific PCR reaction. This can be achieved by carrying out 6 or 24 identical reactions for a 24 or 96 well block, respectively. The reactions should be placed at various positions around the block; in particular, the corner and edge wells should be represented. Thermal cyclers should always be fully serviced and calibrated by the manufacturer to ensure that they are working at optimum efficiency and accuracy. The use of ambient thermal cyclers is not recommended for RAPD analysis owing to the low annealing temperature; however, improvement of reproducibility can sometimes be achieved by carrying out reactions in reduced temperature environments such as cold rooms.

7.4.5 Primer Concentration and Design

Traditional RAPD primers are 10 base pairs in length; however, related techniques employing primers from 5 to 25 bases have been investigated (see Table 7.3). There is no guarantee that a given primer will amplify a given target, and the most suitable primers for an assay will have to be determined experimentally by screening a selection of primers. Figure 7.4 demonstrates that even a single base change at the penultimate 3′ position of the primer can produce highly variable profiles. Commercially available RAPD primers are usually of 60–70% G/C content, which is recommended to optimise reproducibility. If working with A/T-rich genomes, then primers with G or C trimer repeats should be avoided as these sequences will not be common in the genome. Primer concentration will require optimisation for each primer/template combination. Figure 7.5 shows how RAPD profiles are affected by both DNA and primer concentration.

Primers can be used pairwise to increase the number of fragments produced, resulting in novel bands being produced that are not found in profiles generated from the primers when used individually.

7.4.6 Denaturation Time

Yu and Pauls[15] recommend keeping denaturation times to a minimum, with times as short as 5 s (94 °C) being acceptable depending on the thermal cycler

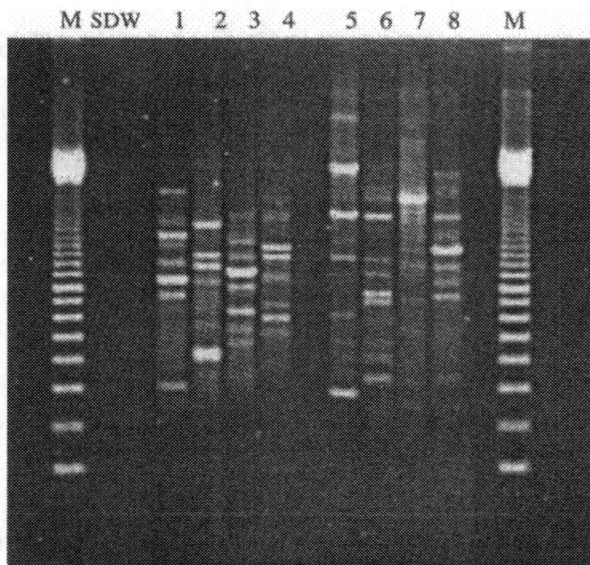

Figure 7.4   *RAPD profile of pUC18 (lanes 1–4) and M13 (lanes 5–8) DNA showing the effect of using primers with a single base change at the penultimate base at the 3' end. Primer sequences are (5'–3') GCAAGGCGA̲T, matching with 100% complementary to a single site on the pUC18 template (lane 1), GCAAGGCGG̲T (lane 2), GCAAGGCGC̲T (lane 3), GCAAGGCGT̲T (lane 4), GCGCGGTAA̲T, matching with 100% complementary to a single site on the M13 template (lane 5), GCGCGGTAG̲T (lane 6), GCGCGGTAC̲T (lane 7) and GCGCGGTAT̲T (lane 8). M represents a 100 bp molecular weight marker*

employed. Minimum denaturation times are thought to reduce the inactivation of the *Taq* DNA polymerase caused by extended incubation at high temperatures.

7.4.7   Annealing Temperature

The annealing temperature most commonly used is 37 °C, however, this can be adjusted from approximately 32 to 40 °C if optimisation requires. Some DNA products may be more temperature sensitive that others; therefore, ideally the chosen optimum annealing temperature should produce identical RAPD profiles at ±1 °C. Figure 7.6 demonstrates the choice of the most suitable primer, based on the effects of annealing temperature, from a selection of four. Primer 2 produces RAPD profiles that are reproducible across a temperature range of 32–42 °C, whereas the other primers produce profiles with bands that appear to be temperature sensitive.

A   B   C   D
[1 2 3 4] [1 2 3 4] M [1 2 3 4] [1 2 3 4]

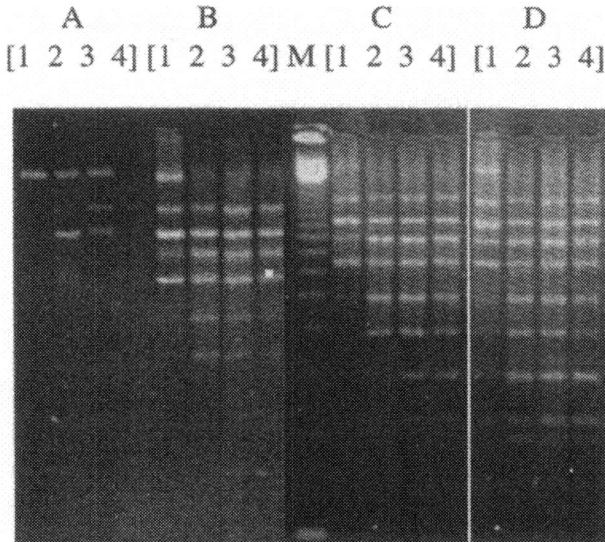

Figure 7.5   *RAPD profiles of* Saccharomyces cerevisiae *using different concentrations of DNA template and primer. Lanes 1–4 represent different concentrations of DNA. 1 is a 20 × dilution, 2 is a 400 × dilution, 3 is a 1000 × dilution and 4 is a 5000 × dilution of the DNA extract. A, B, C and D refer to primer concentrations of 0.1, 0.5, 2.0 and 4.0 µM, respectively. Higher DNA template concentrations (lanes 1) appear to amplify larger products preferentially whereas higher primer concentrations favour the production of smaller fragments in the profile*

7.4.8   Ramp Rate

The rate of temperature change (ramp rate) between the annealing and extension steps of the PCR can also influence the reproducibility of RAPD profiles. The default setting on a thermal cycler is usually set for the fastest transition time possible. Confusingly, different companies quantitate the rate of ramp in different ways, representing it as a °C change per second, seconds change per °C, as a percentage of the default ramp rate or as a transition period from one temperature to the next. Whichever system is used, a reduction of the ramp rate may be required in order to produce reproducible RAPD profiles. This may be especially important when amplifying small genomes as slower ramp conditions, particularly from annealing to extension, can favour the mis-priming events on which the generation of products may be dependent (see Section 7.1).

7.4.9   Extension Time

Although little investigation of this parameter appears to have taken place, Yu and Pauls[15] have shown that when using plant DNA an increase in extension

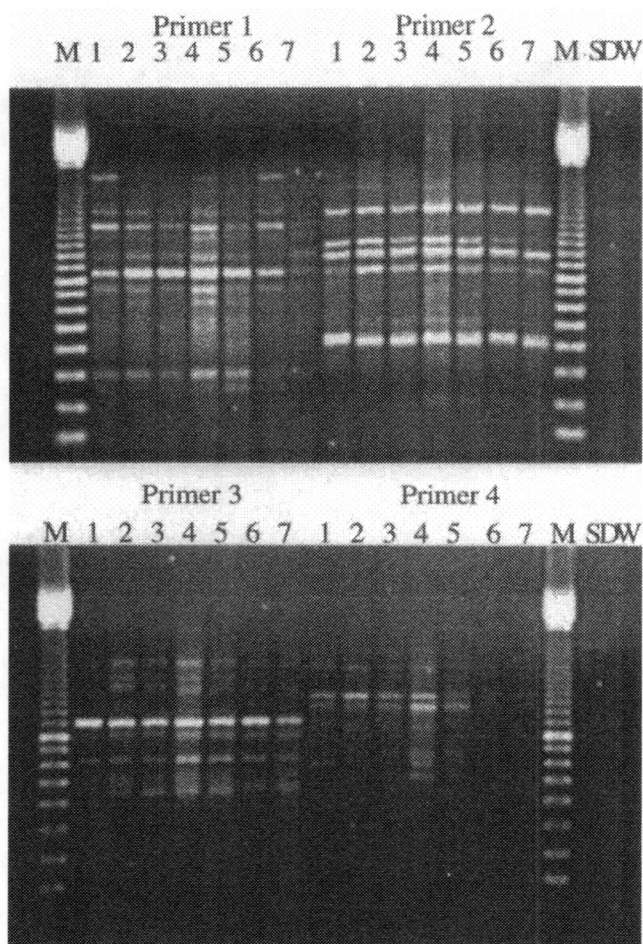

Figure 7.6 *RAPD profile of pUC18 DNA amplified using four primers identical with the exception of a single base change at the 3' end. DNA was amplified with primers of sequence GCAAGGCGAT (1), GCAAGGCGGT (2), GCAAGGCGCT (3) and GCAAGGCGTT (4). The annealing temperatures used were 37°C (lane 1), 32°C (lane 2), 34°C (lane 3), 36°C (lane 4), 38°C (lane 5), 40°C (lane 6) and 42°C (lane 7). M represents a 100 bp molecular weight marker*

time from 5 s to 1 min gives an increase in larger fragments observed. Of the primers they tested, an extension time of 5 s produced PCR products less than 1 kbp, whereas using extension time of 60 s the same primers produced PCR products greater than 3 kbp in size.

7.4.10 Intensities of Amplified Products

Venugopal *et al.*[4] have investigated how RAPD profiles are generated and suggest that the strong signals obtained are a result of two perfectly complementary inverted sequences being present at the primer sites, whereas fainter bands may be due to one primer being mis-matched. However, research in our laboratory, using fully sequenced vectors pUC18 and M13 as templates, has shown that the differences in band intensity commonly observed are obtained even when no 100% complementary sequence exists between the primer and template sequences (Figure 7.4). One primer was designed to anneal to a single 100% complementary binding site on the template. A single base change at the penultimate 3′ base provided the other primers used, where a single 90% complementary sequence match between the primer and template was available. Analysis of the number and sizes of the fragments produced, suggests that the majority of the fragments arose from two priming events with 70% complementarity. Bands of strong intensity are clearly seen in all the profiles obtained, even though two perfect priming sites do not exist on the template.

Bands of varying intensity seen on agarose gels can sometimes be due to co-migrating bands of similar sizes. Employment of electrophoresis techniques such as polyacrylamide can improve the resolution of such bands.

7.4.11 The Addition of PCR Enhancers

As described in detail in Chapter 6, certain chemicals known as enhancers or co-solvents can be added to a PCR reaction to improve the yield of the amplification products. This also applies to RAPD reactions. It should be noted, however, that as a result of increasing product yield, bands that were not previously detectable under non-enhancing conditions can become visible, as seen in Figure 7.7. Care should therefore be taken not to make direct comparisons between profiles generated with and without the addition of enhancers, as additional bands may be falsely interpreted as DNA polymorphisms.

7.4.12 Interpretation of RAPD Profile Data

Polymorphisms are scored on the simple presence or absence of a band of a given size. However, it cannot always be presumed that bands of a similar size will be of homologous sequence and therefore closely related, especially when distant species are examined. To avoid possible anomalies that can occur with the assumption that bands of the same size have the same sequence, Pillay and Kenny[20] and Clark and Lanigan[21] suggest that band homology should be

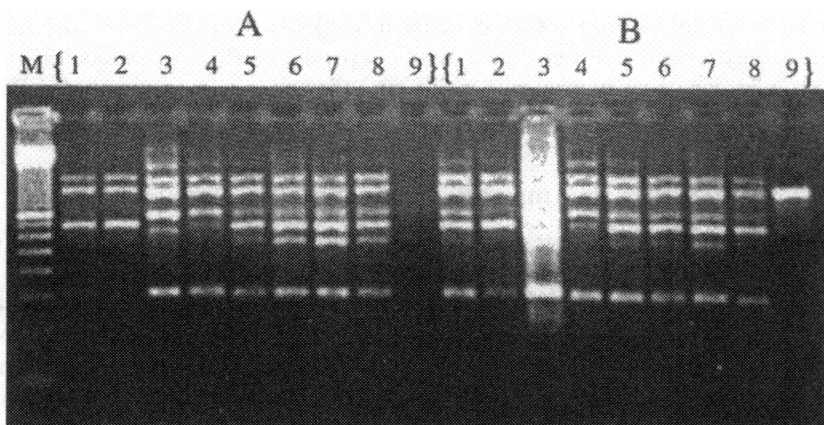

Figure 7.7   *The effect of putative PCR enhancers on the reproducibility of RAPD profiles generated from* Saccharomyces cerevisiae. *A and B represent two different methods of DNA extraction. Enhancers used were 75 µg/ml BSA (lane 2), 200 µg/ml BSA (lane 3), 500 µg/ml BSA (lane 4), 0.1% DMSO (lane 5), 1% DMSO (lane 6), 2.5% DMSO (lane 7), 0.1 M betaine (lane 8) and 1 M betaine (lane 9). The control (lane 1) contains no additives. M is the 100 bp molecular weight marker*

confirmed by Southern blotting and hybridisation. Other techniques such as denaturing gradient gel electrophoresis and temperature sweep gel electrophoresis can be employed to improve resolution and provide more complex profiles. A simple way around this problem is to subject isolates to analysis with a minimum of 10 different primers before relatedness is inferred, thus minimising the affect of coincidental non-homologous bands of equal size. However, it should be noted that RAPD analysis may not be a suitable technique for identification at a higher taxonomic level. This view is confirmed by Clark and Lanigan[21] and Clark,[22] who suggest that profile data from arbitrary primer methods become less predictively informative as phylogenies diverge.

In order to improve the interpretation of data and avoid the error introduced by measuring the *Rf* value by eye, image analysis equipment can be used to compare band sizes objectively. Alternatively, GeneScan equipment with fluorescently labelled primers could be employed to analyse the RAPD profiles.

7.5   Related Technologies

This chapter has been concerned with RAPD profiling; however, there are many other related techniques where no prior sequence knowledge is required (Table 7.3). The majority of the issues discussed in this chapter are also relevant to these technologies and in particular the guidelines set out in Section 7.6 could

Table 7.3 *Techniques related to RAPD profiling*

Name	Description; advantages (+) and disadvantages (−)
AFLP (amplified fragment length polymorphism)	Restriction enzyme digestion of template and ligation of linkers prior to amplification of complete fragments with specific fluorescently labelled primers[23] (+) Enhances the detection of polymorphisms in very closely related isolates or small genomes. Automated detection and fragment sizing when used in association with ABI Sequencer. Improved reproducibility (−) Two- or three-stage process; can be time consuming. Dedicated equipment required
AP-PCR (arbitrary primed PCR)	Longer primer (20mer); initial low stringency followed by high stringency PCR; analysis on polyacrylamide gels; detection by autoradiography[1] (+) Highly complex profile produced (−) Initial low stringency could lead to poor reproducibility. Radioactive labelling used
ASAP (arbitrary signatures from amplification profiles)	Reamplification of previous DAF profile with mini-hairpin or standard oligonucleotides[24] (+) Allows additional screening of amplicons. Hairpin oligonucleotides can improve reproducibility (−) Second amplification required
DAF (DNA amplification fingerprinting)	Shorter primers (5–12mer); analysis on polyacrylamide gels; detection by silver staining. Produces more complex profiles[25] (+) Highly complex profile produced. Stringent primer annealing (~ 50 °C) reduces mis-matching (−) Time consuming detection
RAHM/RAMPO (random amplified hybridisation microsatellites/ microsatellite polymorphisms)	Combination of random amplification, Southern transfer and hybridisation with microsatellite sequence primers[26,27] (+) Enhances polymorphic detection; therefore useful for very closely related isolates. Blots can be rehybridised with different probes (−) Time consuming; reproducibility a concern
RAPD (random amplified polymorphic DNA)	Amplification with 10 base pair primers[1,2] (+) Simple procedure (−) Simple profiles obtained. Low stringency PCR conditions can reduce reproducibility
tecMAAP (template endonuclease cleavage multiple arbitrary amplicon profiling)	Restriction enzyme digestion of template prior to amplification[28] (+) Enhances the detection of polymorphisms in very closely related isolates or small genomes (−) High quality DNA required; samples must be checked to ensure that complete digestion has occurred

increase the validity of these related techniques. Amplified fragment length polymorphism (AFLP) analysis in particular is proving a popular and reproducible profiling technique, with the added benefit of an automated detection system (e.g. ABI sequencer). Originally employed primarily for plants it is increasingly used on a wide variety of microorganisms.

7.6 Concluding Comments

The first description of RAPD profiling in 1990 presented a novel approach to DNA fingerprinting or profiling techniques that was soon adopted around the world in research and analytical laboratories. Following the initial rush of application, investigations into the validity of the technique soon followed, with many questions raised regarding the reproducibility of the technique.[17–20,29] However, it now appears that more sophisticated and considered approaches to primer design, amplification, visualisation and analysis have been sought. In addition, there is a greater understanding of the factors that are critical to the reproducibility of the technique. The advantages of RAPD profiling remain that it is a quick, simple technique, is not expensive, no radio-labelling is involved and no previous knowledge of the target genome sequence is required.

Users should, however, remain aware of the severe limitations of the technique. Both the low specificity and random nature that are inherent to the technique may mean that reproducibility will remain a concern. In 1993, Meunier and Grimont[17] noted that unless standardised, the RAPD method will not enable the constitution of a data bank of profile patterns for identification purposes, which limits the usefulness of data collected. In order to maximise the validity of data generated using this technique, eight guidelines are set out below which could assist in standardising and enhancing, if not the reproducibility, then at least the repeatability of RAPD data.

- Invest time in obtaining good quality, high molecular weight DNA.
- Quantitate DNA template and test for reproducibility of profiles with concentrations of at least a two-fold dilution and a two-fold concentration of the optimum.
- Use a high quality thermal cycler block with highly efficient heat transfer properties, calibrate equipment and check for well-to-well variation.
- Use optimised thermal cycling components, particularly annealing temperature, ramp rate and primer and $MgCl_2$ concentration.
- Always quote all reaction conditions, enhancers and equipment used in publications presenting RAPD data.
- Take maximum precautions to avoid contamination of reactions and carry out appropriate controls to monitor the situation.
- Use the best possible techniques for PCR product separation according to the complexity of the profile required.
- Analyse data from profiles generated with at least six different primers.

7.7 References

1. Welsh, J. and McClelland, M. 1990. Fingerprinting genomes using PCR with arbitrary primers. *Nucleic Acids Res.* **18**: 7213–7218.
2. Williams, J. G. K., Kubelik, A. R., Livak, A. R., Rafalski, J. A. and Tingey, S. V. 1990. DNA polymorphism amplified by arbitrary primers are useful as genetic markers. *Nucleic Acids Res.* **18**: 6531–6535.
3. Caetano-Anollés, G., Bassam, B. J. and Gresshoff, P. M. 1992. Primer–template interactions during DNA amplification fingerprinting with single arbitrary oligo-nucleotides. *Mol. Gen. Genet.* **235**: 157–165.
4. Venugopal, G., Mohapatra, S., Salo, D. and Mohapatra, S. 1993. Multiple mismatch annealing: basis for random amplified polymorphic DNA fingerprinting. *Biochem. Biophy. Res. Commun.* **197**: 1382–1387.
5. Williams, J. G. K., Hanafey, M. K., Rafalski, J. A. and Tingey, S. V. 1993. Genetic analysis using random amplified polymorphic markers. *Methods Enzymol.* **218**: 704–740.
6. Louden, K. W., Burnie, J. P., Coke, A. P. and Matthews, R. C. 1993. Application of polymerase chain reaction to fingerprinting *Aspergillus fumigatus* by random amplification of polymorphic DNA. *J. Clin. Microbiol.* **30**: 1117–1121.
7. Smith, M. L., Bruhn, J. N. and Anderson, J. B. 1992. The fungus *Armillaria bulbosa* is among the largest and oldest living organisms. *Nature* **321**: 428–431.
8. Weising, K., Nybom, H., Wolff, K. and Meyer, W. 1995. DNA Fingerprinting in Plants and Fungi. CRC Press, Boca Raton, FL.
9. Cobb, B. 1997. Optimization of RAPD fingerprinting. In: Fingerprinting Methods Based on Arbitrarily Primed PCR (eds. Micheli, M. R. and Bova, R.), pp. 93–103. Springer, Berlin.
10. Walsh Weller, J. and Reddy, A. 1997. Fluorescent detection and analysis of RAPD amplicons using the ABI PRISM DNA sequencers. In Fingerprinting Methods Based on Arbitrarily Primed PCR (eds. Micheli, M. R. and Bova, R.), pp. 81–92. Springer, Berlin.
11. Dweikat, I and Mackenzie, S. 1997. Denaturing gradient gel electrophoresis for enhanced detection of DNA polymorphisms. In: Fingerprinting Methods Based on Arbitrarily Primed PCR (eds. Micheli, M. R. and Bova, R.), pp. 135–141. Springer, Berlin.
12. Penner, G. A. 1997. Modified temperature sweep gel electrophoresis. In: Finger-printing Methods Based on Arbitrarily Primed PCR (eds. Micheli, M. R. and Bova, R.). Springer, Berlin.
13. Bazzicalupo, M. and Fani, R. 1997. The use of RAPD for generating specific DNA probes for microorganisms. In: Species Diagnostics Protocols: PCR and Other Nucleic Acid Methods (ed. Clapp, J. P.), pp. 155–175. Humana Press, Totowa, N.J.
14. Nei, N. and Li, W. M. 1979. Mathematical model for studying genetic variation in terms of restriction endonucleases. *Proc. Natl. Acad. Sci. USA* **76**: 5269–5273.
15. Yu, K. and Pauls, P. 1994. Optimization of DNA-extraction and PCR procedures for Random Amplified Polymorphic DNA (RAPD) analysis in plants. In: PCR Technology Current Innovations (eds. Griffin, H.G. and Griffin, A.M.), pp. 193–200. CRC Press, Boca Raton, FL.
16. Bassam, B. J., Caetano-Anollés, G. and Gresshoff, P. M. 1992. DNA amplification fingerprinting of bacteria. *Appl. Microbiol. Biotechnol.* **38**: 70–76.
17. Meunier, J.-R. and Grimont, P. A. D. 1993. Factors affecting reproducibility of random amplified polymorphic DNA fingerprinting. *Res. Microbiol.* **144**: 373–379.
18. MacPhearson, J. M., Eckstein, P. E., Scoles, G. J. and Gajadhar, A. A. 1993. Variability of the random amplified polymorphic DNA assay among thermal cyclers, and effects of primer and DNA concentration. *Mol. Cell. Probes* **7**: 293–299.
19. Penner, G. A., Bush, A., Wise, R., Kim, W., Domier, L., Kasha, K., Laroche, A., Scoles, G., Molnar, S. J. and Fedak, G. 1993. Reproducibility of Random Amplified

Polymorphic DNA (RAPD) analysis among laboratories. *PCR Methods Applications* **2**: 341–345.
20. Pillay, M and Kenny, S. T. 1995. Anomalies in direct pair-wise comparisons of RAPD fragments for genetic analysis. *BioTechniques* **19**: 694–698.
21. Clark, A. G. and Lanigan, E. M. S. 1993. Prospects for estimating nucleotide divergence with RAPDs. *Mol. Biol. Evol.* **10**: 1096–1111.
22. Clark, A. G. 1997. Estimating nucleotide divergence with RAPD data. In: Fingerprinting Methods Based on Arbitrarily Primed PCR (eds. Micheli, M. R. and Bova, R.), pp. 219–225. Springer, Berlin.
23. Vos, P., Hogers, R., Bleeker, M., Reijans, M., van de Lee, T., Hornes, M., Frijters, A., Pot, J., Peleman, J., Kuiper, M. and Zabeau, M. 1995. AFLP: a new technique for DNA fingerprinting. *Nucleic Acids Res.* **23**: 4407–4414.
24. Caetano-Anollés, G. and Gresshoff, P. M. 1996. Generation of sequence signatures from DNA amplification fingerprints with mini-hairpin and microsatellite primers. *BioTechniques* **20**: 1044–1056.
25. Caetano-Anollés, G., Bassam, B. J. and Gresshoff, P. M. 1991. DNA amplification fingerprinting using very short arbitrary oligonucleotide primers. *BioTechnology* **9**: 553–557.
26. Cifarelli, R. A., Gallitelli, M. and Cellini, F. 1995. Random amplified hybridization microsatellites (RAHM): isolation of a new class of microsatellite-containing DNA clones. *Nucleic Acids Res.* **23**: 3802–3803.
27. Richardson, T., Cato, S., Ramser, J., Kahl, G. and Weising, K. 1995. Hybridization of microsatellites to RAPD: a new source of polymorphic markers. *Nucleic Acids Res.* **23**: 3798–3799.
28. Caetano-Anollés, G., Bassam, B. J. and Gresshoff, P. M. 1993. Enhanced detection of polymorphic DNA by multiple arbitrary amplicon profiling of endonuclease digested DNA; identification of markers linked to the supernodulation locus in soyabean. *Mol. Gen. Genet.* **241**: 57–64.
29. Ellsworth, D. L., Rittenhouse, K. D. and Honeycutt, R. L. 1993. Artifactual variation in randomly amplified polymorphic DNA banding patterns. *BioTechniques* **14**: 214–217.

CHAPTER 8

Development of Multiplex PCR

JO SHORT AND JIM THOMSON

8.1 Introduction

Many experimental approaches require analysis of a variety of DNA sequences, necessitating multiple polymerase chain reactions to be performed on the same or related templates. Considerable savings of time, effort and reagent costs can be achieved by simultaneously amplifying multiple sequences in a single reaction, a process referred to as multiplex PCR.[1] Whilst there is no theoretical limit to the number of loci which can be amplified simultaneously, there are a number of practical constraints to be considered which place limits on the technique.[2]

Firstly, as the number of loci increase it becomes increasingly difficult to balance the optimum PCR reaction conditions for all loci. Even small variations in the efficiency of PCR amplification across the loci can lead to marked differences in product yield, and this may present difficulties in the detection and interpretation of all the loci in a multiplex. Another limitation is the increasing chance of non-specific amplification between non-paired primers. As the number of primers in the reaction increases, the permutations of primer pairs which may interact non-specifically increases significantly. The amplification of significant amounts of non-specific product may hinder or prevent a clear interpretation of the results of some or all of the desired loci. Finally, the detection system employed must be able to identify unambiguously the products from every locus amplified in the multiplex PCR (mPCR). As the number of loci increases, the differentiation of these products can become more difficult, irrespective of the detection system is employed.

Multiplex PCR (mPCR) is generally divided into two groups:

- Linked mPCR, where different regions of the same gene or genome are amplified (Figure 8.1)
- Non-linked mPCR, where sequences located on unrelated but co-extracted or premixed genomes are amplified (Figure 8.2).[3]

Multiplex PCR is an extremely flexible technique and has numerous applications such as pathogen identification, gender screening, linkage analysis,

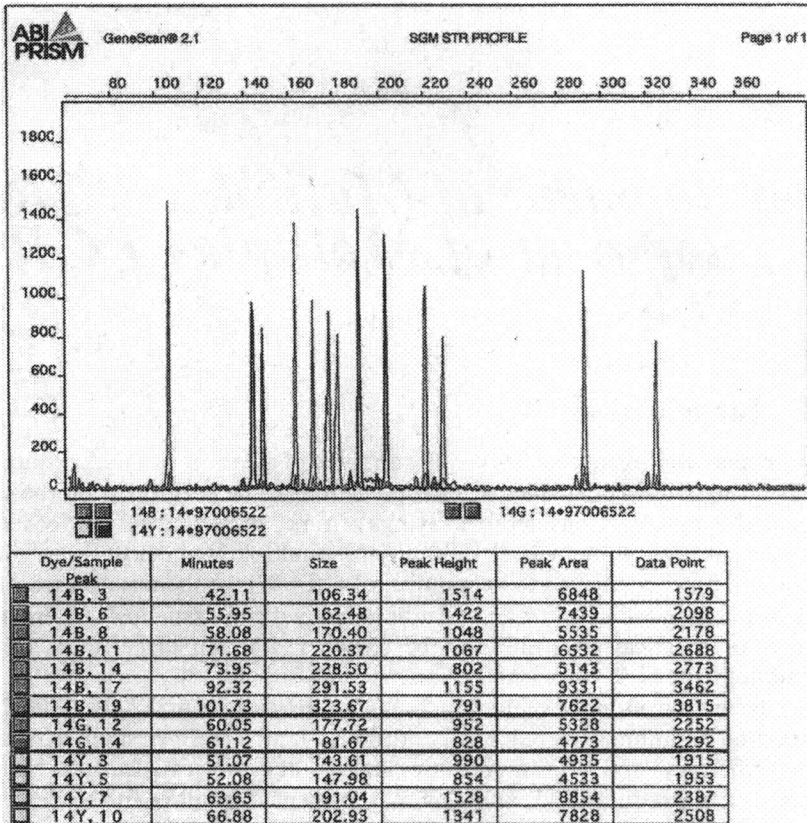

Dye/Sample Peak	Minutes	Size	Peak Height	Peak Area	Data Point
14B, 3	42.11	106.34	1514	6848	1579
14B, 6	55.95	162.48	1422	7439	2098
14B, 8	58.08	170.40	1048	5535	2178
14B, 11	71.68	220.37	1067	6532	2688
14B, 14	73.95	228.50	802	5143	2773
14B, 17	92.32	291.53	1155	9331	3462
14B, 19	101.73	323.67	791	7622	3815
14G, 12	60.05	177.72	952	5328	2252
14G, 14	61.12	181.67	828	4773	2292
14Y, 3	51.07	143.61	990	4935	1915
14Y, 5	52.08	147.98	854	4533	1953
14Y, 7	63.65	191.04	1528	8854	2387
14Y, 10	66.88	202.93	1341	7828	2508

Figure 8.1 *Example of a linked mPCR reaction. Six short tandem repeat (STR) loci were multiplexed along with the Amelogenin XY locus. One of each primer pair was labelled with one of three different fluorescent dyes (HEX, 6-FAM, TET). Products were detected by electrophoresis on a PE-ABI 377 DNA Sequencer which detects the fluorescent dyes as they pass a fixed point scanning laser. The loci detected are (l to r) Amelogenin (blue), vWA (black), TH01 (blue), D8S1179 (green), FGA (black), D21S11 (blue) and D18S51 (blue). These loci constitute the 'second generation multiplex' (SGM) developed specifically for the purpose of human identification. Molecular sizes of the detected fragments, expressed in base pairs, are shown in the table*

forensic studies[4] and genetic disease diagnosis.[5] It can be an end point analysis in itself or be a preliminary stage prior to techniques such as sequencing, hybridisation and restriction digest analysis.[5-7] From a theoretical view point it allows the influence of such conditions as cycling parameters, enhancers and inhibitors to be assessed using a range of targets simultaneously.

The general advantages of mPCR, in addition to the savings in time, effort and reagents costs, include the removal of tube to tube variation when using amplification controls. It can also give an indication of template quantity, when used in conjunction with target mimics,[8] and quality. For example, a reduction

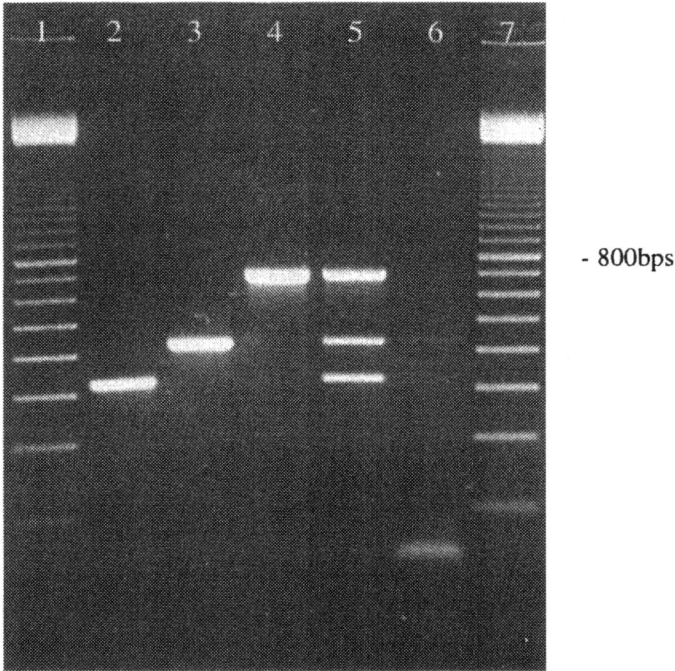

Figure 8.2 *Non-linked mPCR for the simultaneous detection of* Escherichia coli *O157,* Salmonella typhimurium *and* Listeria monocytogenes. *Lanes 1 and 7, 100 bp molecular weight marker. Lane 2,* E. coli *(O157) DNA. Lane 3,* S. typhimurium *DNA. Lane 4,* L. monocytogenes *DNA. Lane 5, multiplex PCR system. Lane 6, negative control*

in the generation of longer target sequences whilst maintaining amplification of shorter targets implies template degradation, whilst a decrease in amplification of abundant control sequences implies amplification inhibition.

The major disadvantage of mPCR is the initial effort required in both theoretical and experimental design, as discussed later (Sections 8.2 and 8.3). It is worth considering early in the design of a multiplex its analytical benefits versus the development time and costs. The more primer sets added to the reaction, the more difficult it becomes to ensure effective amplification and differentiation of all DNA fragments. Simple approaches may prove the most beneficial; for example, the development of two triplex reactions rather than a single sextuplex may prove a more suitable solution.

8.2 Design of a Multiplex PCR

8.2.1 Choosing Targets

The regions chosen for amplification will essentially be determined by the nature of the analysis, e.g. microbial identification assays may target species or strain specific variations or toxin genes, forensic assays will distinguish individual variations at highly polymorphic loci. If internal controls are required, their source must also be considered carefully. Common targets are housekeeping genes such as cytochrome *b* or ribosomal RNA genes. In some instances it may be important to match the DNA source of the locus to be analysed and the internal control locus.

8.2.2   Position of Primers

Several factors should be considered when deciding where to position primers for a mPCR system. First, the sequence of the flanking regions of the target site may impose constraints, as discussed in Section 8.2.3. This may be either that insufficient flanking sequence is available to give any flexibility in primer position, or that flanking regions contain sequences unsuitable for primer sites, such as non-unique or repetitive sequences.

The detection of the multiplexed products must also be considered. If gel electrophoresis is to be used as the end point detection system it is important to position primers such that the fragments may be easily separated by size, although the size range should not be so great that they cannot be separated on a single gel. Other detection systems may not be dependent on difference in fragment size to distinguish products, allowing identical or very similar sized products to be identified. For example, the use of fluorescent dye based detection systems, even if electrophoresis based, may allow two or more co-migrating products to be analysed, provided they are labelled with different coloured fluorescent dyes. Other systems, such as solid phase capture systems, may impose no significant restrictions on product sizes and so allow greater freedom in primer position and choice (Section 8.3).

The different efficiency of amplification of loci within a multiplex is also an issue when positioning mPCR primers. A large size range in the expected products may well be manifested by preferential amplification of smaller PCR products over larger ones. In extreme cases this can result in the failure of one or more loci to amplify. When developing linked mPCRs it is worth noting that overlapping targets will result in a mix of nested products rather than the two desired amplicons, and therefore such positioning should be avoided. When high specificity is required it is important to choose priming sites from variable regions, allowing a site unique to the target of interest to be chosen. The more sequence data that are available for potential priming sites, the easier it is to develop compatible primer sets.

8.2.3   Primer Design

Primer design for multiplex PCR is governed by the same rules as for single PCR amplifications, but additional factors must be considered to ensure that a compatible primer set is obtained. An important first step in designing a multiplex reaction is to ensure that the predicted melting temperatures (T_m), the temperature at which 50% of the template annealing site is in duplex with the primer, for each primer set are similar In addition, each primer within a primer set should have a similar T_m. Melting temperature values are best calculated using one of the commercial primer design packages currently available, e.g. Oligo, Primers 1.2. If it is not possible to calculate predicted T_m values, empirical indicators are that the primers should be as uniform in both length and GC content as possible. GC distribution along the length of the fragments should also be uniform, with no high-melting temperature regions.

It should be noted that calculated T_m values should not be taken as being completely reliable indicators of primer behaviour. Rather, they should be seen as a starting point from which optimisation of performance can proceed. If one locus in a multiplex system is later seen to result in low product yields, or is identified as a source of artefact products, it may be that the true T_m is significantly different from that calculated and it will be necessary therefore to redesign one or both of the primers to overcome this.

Primers should also be checked to ensure low homology, particularly at the 3′ end, both within sets and between sets, to limit the generation of primer dimers and other artefacts. Information of this kind can also be obtained from commercially available primer design software packages. A mPCR may be generated by combining published primer sets, specific to the regions of interest, but designing sets specifically to work together gives greater flexibility and can avoid many problems.

8.2.4 Initial PCR Development

PCR conditions should be optimised empirically for each primer set individually to establish their sensitivity and specificity. Particular attention should be paid to optimal annealing temperature and time using standard buffering conditions. Primer sets that do not function well under standard conditions are best replaced.

Having established optimal conditions for each primer set, a single central set should be chosen and sets sequentially added to this to establish initially a duplex, then a triplex and so on. Clearly, the optimum conditions for each primer set may vary, so it is important to continue an empirical process of optimisation as more primer sets are added. The aim is to develop a set of conditions that represent an acceptable compromise as far as the amplification efficiency of the component loci are concerned. An overall decrease in sensitivity of the reaction may be observed on development of the multiplex reaction. This is in part due to the sub-optimal amplification conditions and also, in the case of non-linked mPCR, may be due to an overall increase in non-specific nucleic acid and additional cellular debris found within the reaction mix.

To obtain uniform amplification of all loci within a multiplex, variation in primer concentrations between loci may be required. This is particularly evident where there are variable target copy numbers, but also if the reaction contains primer sets with highly variable primer/target annealing efficiencies. Such variations can be highlighted by carrying out cycle by cycle analysis, i.e. the removal of an aliquot of the reaction mix after each PCR cycle and subsequent

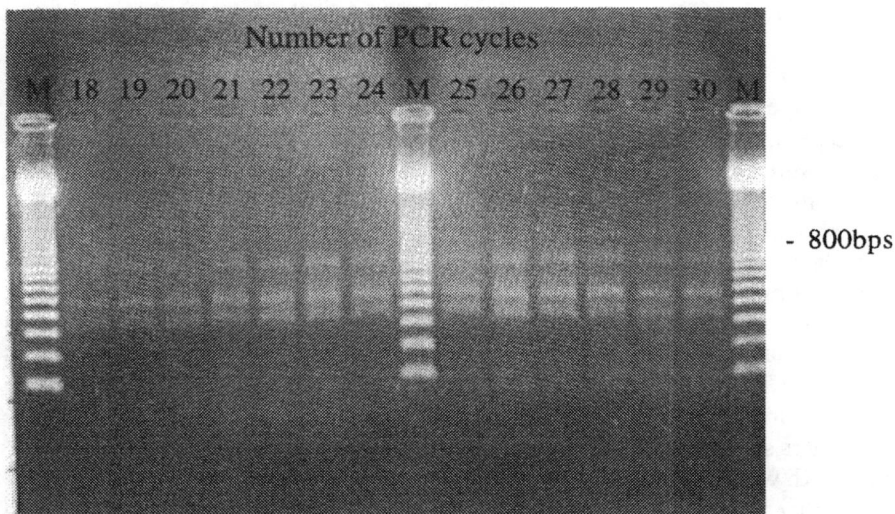

Figure 8.3 *Cycle-by-cycle analysis of a multiplex amplification reaction. 3 μl aliquots were removed at each cycle from a large-volume PCR. M, 100 bp molecular weight marker. Lanes 18–30 indicate the number of PCR cycles carried out*

analysis. Any differences in the yield of unequally amplified fragments may be enhanced with each cycle and therefore become apparent. Although labour intensive, this method can highlight problematic primer sets and it is also useful in establishing a minimum cycle number (Figure 8.3).

Other factors which may require consideration are buffer composition (particularly Mg^{2+} concentration), template quantity and quality, polymerase type and source, reaction volume, effect of oil overlay, performance of and transferral between thermal cyclers, benefit of hot start or touchdown PCR, number of cycles and temperature, time and ramp rates of all steps within the PCR cycle, many of which are discussed further below.

8.2.5 Reaction Components

Owing to the compromises that are necessarily made to reaction parameters to produce simultaneous amplification of multiple sequences, the resulting mPCR may lack the robustness required for its desired application. Careful manipulation of reaction components can result in an increase in robustness, aiding the repeatability, reproducibility and sensitivity of the reaction.

Increasing magnesium ion, nucleotide and enzyme concentration may be necessary on increasing the number of targets within the reaction, considering that the total quantity of amplified product generated within a mPCR is usually greater than that generated in a uniplex reaction. Magnesium ion concentration, in particular, should be titrated carefully as the resulting decrease in specificity can outweigh the benefits of increased amplification (Figure 8.4). Owing to the

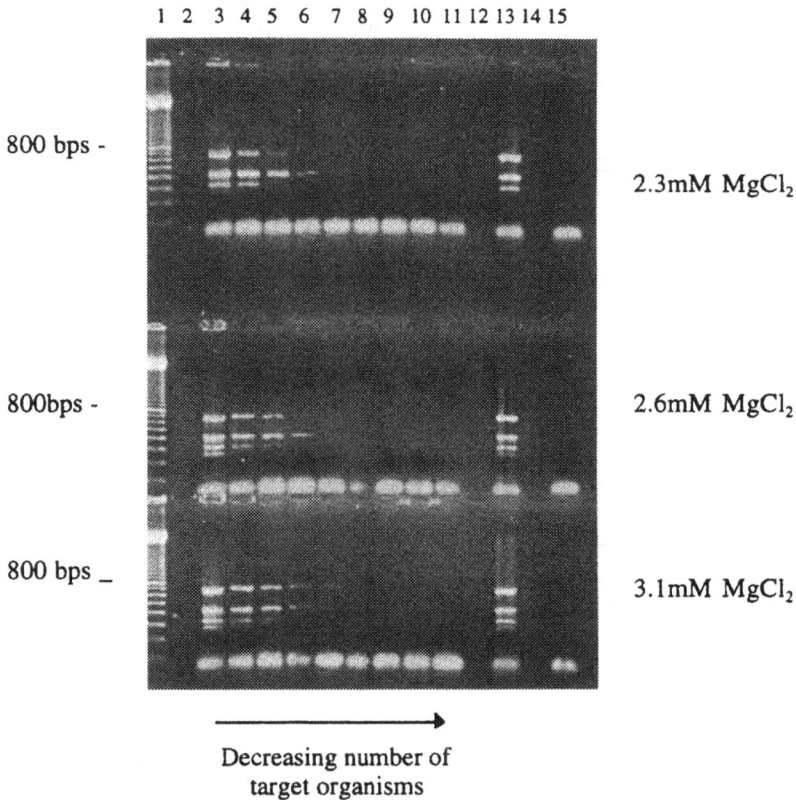

Figure 8.4 *Effect of increasing magnesium chloride concentration, from 2.3 to 3.1 mM, on the sensitivity of a multiplex reaction. Lane 1, 100 bp molecular weight marker. Lanes 2, 12 and 14, blank. Lanes 3–11, 10^6 cfu–0 cfu of each organism. Lane 13, positive control. Lane 15, negative control*

requirement of the polymerase enzyme for free magnesium ions and the chelation of magnesium ions by nucleotides, an increase in nucleotide concentration without a related increase in magnesium ion concentration can rapidly inhibit the PCR. We have found that a free magnesium ion concentration in excess of 1 mM is desirable. Optimisation of buffering conditions needs to be carried out with reference to each specific multiplex reaction as changes, such as the use of adjuvants, can give conflicting results.[9]

The quality and quantity of the template can affect the robustness of the most carefully designed mPCR. Owing to the often sub-optimal nature of the reaction, the presence of inhibitory agents within the nucleic acid extract is tolerated poorly, leading to limited amplification, or indeed complete loss of desired products. It may prove necessary to carry out an additional 'clean up' or dilution step during sample preparation when the extract is to be used in a mPCR. When using non-linked PCR to amplify template DNA isolated from different organisms, it may be necessary to investigate the relative efficiency of DNA recovery from those organisms. An optimum DNA extraction technique will isolate DNA with equal efficiency from all the target organisms.

8.2.6   Cycling Parameters

As previously stated, the annealing time and temperature of the reaction are very important and need to be considered early in the development of the reaction. Small changes within these parameters can have a significant affect on the outcome of the overall reaction. As the number of amplification products increases, there must be a corresponding increase in extension time. This ensures the full extension of targets with longer length products and the efficient amplification of primer sets with sub-optimal annealing to the target. The minimum number of cycles that allows easy detection of all fragments of interest should be used. Additional cycles can lead to difficulties in interpretation of results, especially if the reaction has been designed to be semi-quantitative.

When optimising reaction components and cycling parameters, as when initially designing a mPCR, it is important to stress the adoption of an empirical approach. Time invested in understanding and optimising all the steps of the PCR cycle will certainly result in a more efficient and robust mPCR.

8.2.7   Untemplated Nucleotide Addition

A well documented feature of *Taq* DNA polymerase is its tendency to add an additional untemplated nucleotide, usually an adenine, to the 3' end of PCR products. This can result in a mixture of products for any one locus, some of which have an additional nucleotide and are therefore one base longer than the rest. This can cause difficulties in interpretation, especially in techniques such as the STR systems employed in forensic analysis, where single base pair resolution of products is required (Figure 8.5).

In multiplex systems, which are necessarily carried out under conditions which are sub-optimal for some loci, but a working compromise for all, it is easier to optimise for the addition of the extra nucleotide, rather then its non-addition. This means that, ideally, all products should be the so-called '*n* + 1' product (with the extra base) rather than the '*n*' product.

The means by which the additional base addition is encouraged is by the incorporation of an extended final extension hold at the end of the PCR reaction. Classically, this has been carried out at the optimal temperature for DNA *Taq* polymerase extension (72 °C), although it has been reported that a 60 °C final extension step encourages more complete addition.[10] Once again, an empirical approach to optimisation is recommended, as different systems are likely to display different efficiencies of non-templated addition.

If particular loci within a multiplex system are particularly prone to split into '*n*' and '*n* + 1' peaks it may be necessary to modify the primer sequences. Specifically, the 5' end of the unlabelled primer (if a fluorescent detection system is employed) is instrumental in controlling the efficiency of this addition. It has

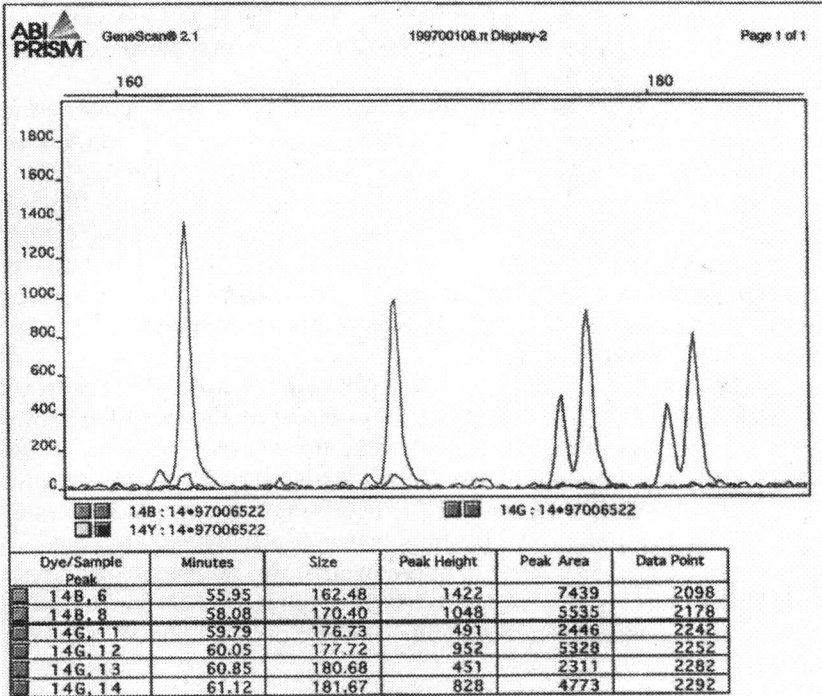

Dye/Sample Peak	Minutes	Size	Peak Height	Peak Area	Data Point
14B, 6	55.95	162.48	1422	7439	2098
14B, 8	58.08	170.40	1048	5535	2178
14G, 11	59.79	176.73	491	2446	2242
14G, 12	60.05	177.72	952	5328	2252
14G, 13	60.85	180.68	451	2311	2282
14G, 14	61.12	181.67	828	4773	2292

Figure 8.5 *Electropherogram of two STR loci (at expanded scale) showing the effect of differential efficiency of untemplated nucleotide addition. The locus on the left (TH01) shows almost complete addition of a non-templated nucleotide (so called 'n + 1' peaks). Only residual 'n' peaks are seen 1 bp smaller than the major allele peak. The locus on the right (D8S1179) shows incomplete addition and very significant 'n' peaks, which may hinder interpretation if genuine single base pair differences between alleles are encountered. The problem may be overcome by modification of primer sequences or the incorporation of an optimised final extension step to encourage non-templated addition*

been demonstrated that primers designed with a purine as the 5′ terminal nucleotide will result in far greater non-templated adenine addition and so drive the product towards the '$n + 1$' form.[10]

8.2.8 Overcoming Mis-priming Events

Mis-priming events, resulting in the generation of non-specific amplification products, not only reduce the specificity of the PCR but also affect the sensitivity, repeatability and reproducibility of the reaction. Their occurrence, and that of other artefacts, increases as more primer sets are added to the reaction. Several methods for avoiding or overcoming such problems have been developed. Commonly, increasing the annealing temperature is used to reduce artefacts, but this must be balanced against the reduced yield of desired PCR products. This is a particularly fine balance if the optimal annealing temperatures of the constituent loci are not the same for all loci.

Hot-start PCR is a valuable technique for reducing primer–dimer and other non-specific primer interactions.[11] A number of methods have been developed for hot-start PCR, many of which are commercially available. The most convenient method for hot-start PCR currently appears to be the use of Amplitaq Gold®, which is a modified *Taq* DNA polymerase that is inactive until it is heated to 95 °C. This allows all reaction components to be mixed at room temperature without the danger of non-specific polymerase-mediated activity. The reaction is then initiated by an activation step at 95 °C before the commencement of thermal cycling.

Touchdown PCR has also been successfully used to reduce the occurrence of non-specific products.[12] The technique involves setting the annealing temperature of the reaction artificially high for the initial cycles and then gradually reducing the set temperature over the following cycles to the empirically determined optimum. The initial high temperature may limit the generation of product but it ensures that the products generated are the desired ones. These products then act as templates in subsequent cycles. Reducing the annealing temperature improves the efficiency of the reaction and, given the increased concentration of the correct target, the occurrence of non-specific products is greatly reduced.

Even in an optimised multiplex system, artefact products may occur. It is helpful to have as much knowledge as possible about these artefact products, as it is likely they will appear in a relatively predictable fashion. Consequently, any artefacts identified should be logged.

8.3   Detection Strategies

8.3.1   Agarose Gel Electrophoresis

Before selecting an electrophoretic format, it is important to consider both the size range of the products to be separated and the required resolution. Electrophoresis separates products by size which is determined by comparing the DNA fragments to those of a molecular weight marker, often run in an adjacent well.

Agarose gel electrophoresis is the most common format employed for the analysis of PCR products and it can equally be applied to mPCR. We have found for the separation of fragments differing in size by > 25 base pairs, a 3% gel of commonly available agaroses such as NuSieve (FMC Bioproducts) functions very well. High sieving agaroses, such as MetaPhor (FMC Bioproducts), may be used in circumstances where higher resolution is required. For example, it is possible to resolve products differing by as little as 4 bp on a 4% MetaPhor agarose gel.

8.3.2 Polyacrylamide Gel Electrophoresis

Native or denaturing polyacrylamide gels can also be used to detect the products of mPCR. Many standard PAGE formats are suitable for such analysis. Detection of products may be by silver staining, ethidium bromide or other DNA-specific staining or a wide variety of other methodologies, none of which are specific for multiplex PCR analysis.

In situations where very high resolution is required to distinguish between DNA fragments of a very similar size, it may be necessary to consider a sequencing gel format consisting of an ultrathin gel (0.2–0.4 mm) of 30 cm or more, run under highly denaturing conditions for example at 8 M urea, at elevated temperatures. This gel format is also employed when using an automated analysis platform such as the PE ABI DNA Sequencer. This allows automatic detection and size estimation of PCR products labelled with fluorescent dyes linked to one of each primer pair. The use of four different fluorescent dyes allows simultaneous detection of overlapping target size ranges and the apparatus is capable of single base pair resolution.

8.3.3 Solid Phase Capture

The capture of the amplified fragments of interest onto a solid support, such as nylon or cellulose filters, the surface of a microtitre plate or column packed microbeads via a target-specific nucleic acid probe, can overcome the problems associated with electrophoresis. If the capture probe used is homologous to a region within the amplified product, as opposed to the primer site itself, the detection of non-specific amplification products will also be avoided.

Detection of the bound product may be through the use of a primer labelled with a hapten such as biotin, which can be detected subsequently using a specific antibody enzyme conjugate such as alkaline phosphatase or horseradish peroxidase or with a directly fluorescent molecule such as fluorescein. Distinction between multiple products is difficult if using labelled primers, but this problem may be overcome using secondary probes each specific to an individual product and labelled distinctly, e.g. with a distinct fluorophore.

As well as overcoming many of the problems related to electrophoresis, solid phase detection systems offer great potential in terms of automation and therefore high throughput of samples. However, it should be noted that they are demanding in terms of development time and require extensive evaluation and validation before routine use.

8.4 Summary

Multiplex PCR is a valuable tool for both research and diagnostic PCR applications. It can save time and resources for the analyst and the laboratory as a whole. This chapter has introduced the concepts of multiplex PCR and has reviewed the considerations which must be made when designing a successful mPCR system.

In summary, two points should be stressed. First, the importance of good primer design. Whilst it may be tempting simply to take published primer sequences and combine them, such a policy is likely to give a sub-optimal and unstable multiplex system. Time spent to ensure that primer sets are compatible in terms of position and sequence is a sound investment.

Further to this is the importance of empirical optimisation. An optimised mPCR system cannot be designed on paper. It will always be necessary to consider individual parameters and to optimise these within the framework of the developing system. However, it is also important to weigh the time spent on optimisation against the final demands on the system; a diagnostic service processing thousands of samples may justify greater resources given to optimisation than a small research project.

8.5   References

1. Chamberlain, J. S., Gibbs, R. A., Ranier, J. E., Nguyen, P. N. and Caskey, C. T. 1988. Deletion screening of the Duchenne muscular dystrophy locus via multiplex DNA amplification. *Nucleic Acids Res.* **16**: 11141–11156.
2. Chamberlain, J. S. and Chamberlain, J. R. 1994. Optimisation of multiplex PCRs. In: PCR, Polymerase Chain Reaction (eds. Mullis, K. B., Ferre, F. and Gibbs, R. A.), pp. 38–46. Birkhäuser, Boston.
3. Pearson, B. M. and McKee, R. A. 1992. Rapid identification of *Saccharomyces cerevisiae*, *Zygosaccharomyces bailii* and *Zygosaccharomyces rouxii*. *Int. J. Food Microbiol.* **16**: 63–67.
4. Kimpton, C. P., Oldroyd, N. J., Watson, S. K., Frazier, R. R. E., Johnson, P. E., Millican, E. S., Urquhart, A., Sparkes, B. L. and Gill, P. 1996. Validation of a highly discriminating multiplex short tandem repeat amplification system for individual identification. *Electrophoresis* **17**: 1283–1293.
5. Cebula, T. A., Payne, W. L. and Feng, P. 1995. Simultaneous identification of strains of *E. coli* serotype O157:H7 and their shiga-like toxin type by mismatched amplification mutation assay — multiplex PCR. *J. Clin. Microbiol.* **33**: 248–250.
6. van der Vliet, G. M., Hermans, C. J. and Klatser, P. R. 1993. Simple colorimetric microtitre plate hybridisation assay for detection of amplified *Mycobacterium leprae* DNA. *J. Clin. Microbiol.* **31**: 665–670.
7. Cuppens, H., Buyse, I., Baens, M., Maynen, P. and Cassiman, J.-J. 1992. Simultaneous screening for 11 mutations in the cystic fibrosis transmembrane conductance regulator gene by multiplex amplification and reverse dot blot. *Mol. Cell. Probes* **6**: 33–39.
8. Gilliland, G., Perrin, S. and Bunn, H. F. 1990. Competitive PCR for quantitation of mRNA. In: PCR Protocols — A Guide to Methods and Applications (eds. Innis, M. A., Gelfand, D. H., Sninsky, J. J. and White, T. J.), pp. 60–69. Academic Press, San Diego, CA.
9. Henegariu, O., Heerema, N. A., Dlouhy, S. R., Vance, G. H. and Vogt, P. H. 1997. Multiplex PCR: critical parameters and step-by-step protocol. *BioTechniques* **23**: 504–511.
10. Hu, G. 1993. DNA polymerase controlled addition of non-templated extra nucleotides to the 3′ end of a DNA fragment. *DNA Cell Biol.* **12**: 763–770.
11. D'Aquila, R. T., Bechtel, L. J., Videler, J. A., Eron, J. J., Gorczyca, P. and Kaplan, J. C. 1991. Maximizing sensitivity and specificity of PCR by preamplification heating. *Nucleic Acids Res.* **19**: 3749.
12. Mellersh, C. and Sampson, J. 1993. Simplifying detection of microsatellite length polymorphisms. *BioTechniques* **15**: 582–584.

CHAPTER 9

Membrane Hybridisation

JOHANNE H. CORNETT, JASON SAWYER AND
DELLA SHANAHAN

9.1 Introduction

The hybridisation analysis of membrane (filter) immobilised DNA is one of the
major analytical tools routinely utilised in molecular biology. The procedure is
fundamental to techniques such as Southern, Northern, colony and dot blotting
which have been exploited in a variety of identification, detection and semi-
quantitative analyses. For example, Southern blotting forms the basis of the
RFLP methodology used in DNA forensics,[1] and slot blot analysis can be
employed to detect and, in some cases, quantify adulteration of food products.[2]

In order to design valid and accurate methods using membrane hybridisation
it is useful to understand the basic principles of the technique. This chapter will
present the fundamentals of nucleic hybridisation and attempt to use this
information to provide a basis for experimental design, development and
validation of membrane hybridisation. A basic understanding of nucleic acid
structure is assumed. This chapter will concentrate on the application of
membrane hybridisation for DNA analysis.

9.1.1 Hybridisation Theory

Nucleic acid hybridisation or reassociation is a process in which complementary
single-stranded nucleic acids anneal to form a double-stranded duplex. The
process is based on the formation of hydrogen bonds between the two strands of
the DNA:DNA, DNA:RNA or RNA:RNA duplexes. The hydrogen bonds
form so that adenine specifically pairs with thymine (or uridine in RNA) and
guanine with cytosine. Hydrogen bonds can be disrupted ('denaturation' or
'melting' of the duplex) or reformed ('renaturation' or 'reannealing' of single
strands to form a double-stranded nucleic acid) repeatedly by simple procedures
without significantly altering the molecular integrity of either strand. Thus the
reannealing reaction is reversible, predictable and controllable. It is important
to realise, however, that reannealing is not simply the reverse of denaturation.
Denaturation can be viewed as a diffusional process similar to simple unwinding

whereas reannealing consists of two processes: (i) nucleation (the specific recognition between two short, complementary segments on each strand), and (ii) zippering (the rapid formation of successive base pairs starting at the nucleation sites).

Membrane hybridisation involves the immobilisation of denatured DNA onto an inert solid support (for example, nitrocellulose or nylon membranes), in such a way that self-annealing is prevented. The solid supported, single-stranded nucleic acid is subsequently incubated in a solution containing a labelled nucleic acid probe. During this time, the probe hybridises to complementary target sequences on the membrane. Detection and localisation of the resultant probe–DNA complexes on the membrane are facilitated by the probe reporter moiety. In general, the term 'hybridisation' refers to the reaction of a probe with its target, as opposed to renaturation and reannealing, which are more general terms describing the formation of a double-stranded duplex from single strands.

9.1.2 Kinetics of Hybridisation

Nucleic acid hybridisation depends on the random collision of two complementary single-stranded sequences. The rate at which hybridisation occurs is determined by the concentration of the reacting species and by the rate constants for hybrid formation (which are determined by experimental conditions).

9.1.2.1 Hybridisation with a Single Membrane-bound Sequence

To illustrate the kinetics of membrane hybridisation, the example of a denatured DNA probe hybridising to a membrane-bound target will be examined. In this case, the membrane-bound nucleic acid consists of only one (the target) sequence. During hybridisation, two competing reactions occur. The two complementary strands of the denatured DNA probe can either reassociate in solution to form a double-stranded probe, or hybridise to the complementary target sequence of the membrane-bound DNA to form a probe–target DNA complex. The rate of disappearance of single strands from the hybridisation solution may be expressed by eqn. (1).[3]

$$\frac{-\mathrm{d}[C_s]}{\mathrm{d}t} = k_1[C_f][C_s] + k_2[C_s]^2 \tag{1}$$

Here C_s is the concentration of nucleic acid probe in solution, k_1 is the rate constant for the hybridisation reaction on the membrane, C_f is the concentration of membrane-bound sequences and k_2 is the rate constant for strand reassociation in solution. A large number of experimental variables affect the rate constants and these are examined in Section 9.3. The effect of probe reassociation is thought to be generally underestimated and can lead to problems. To overcome these, conditions which facilitate diffusion of the probe to the filter and favour hybridisation over reassociation should be chosen. These include using a short, preferably single-stranded, probe at a low concentration in solution, a small reaction volume and a high reaction temperature.

It has been shown experimentally[3] that the initial rate of hybridisation is proportional to the concentration of the probe, as eqn. (1) predicts. However, this equation does not accurately describe the initial hybridisation rate in relation to the concentration of the membrane-bound target DNA. At low concentrations the hybridisation rate is proportional to the amount of target DNA, but does not increase linearly as the concentration of target increases. This is because, at high target concentrations, hybridisation occurs so fast that the solution surrounding the membrane becomes depleted of probe and the overall reaction is limited by probe diffusion. Two types of membrane hybridisation can therefore be distinguished: diffusion-limited and nucleation-limited hybridisation. In the former case, the hybridisation rate will be increased by factors which increase diffusion of the probe to the filter. These include using a small probe, a high incubation temperature, a low reaction volume and agitating the reaction vessel to bring a greater percentage of the probe into contact with the membrane surface. In the latter, the hybridisation rate will be increased by factors that enhance hybrid formation, such as high salt concentration and low temperatures.

9.2 The Various Stages of Membrane Hybridisation

Filter hybridisation experiments can be divided into the following steps, each of which will be briefly outlined:

- DNA extraction
- DNA electrophoresis (when appropriate, i.e. for RFLP analysis and not dot/slot blot-based assays)
- Immobilisation of DNA onto the membrane
- Preparation and labelling of nucleic acid probes
- Hybridisation
- Detection

A flow diagram (Figure 9.1) illustrates the steps involved in a typical slot blot analysis.

9.2.1 DNA Extraction

DNA must be extracted from the sample of interest prior to electrophoresis or immobilisation on a membrane. Appropriate DNA extraction methods vary greatly, depending on the sample. The principles underlying DNA extraction protocols, and the validation of DNA extraction procedures, are discussed in detail in Chapter 3 and will not be covered here.

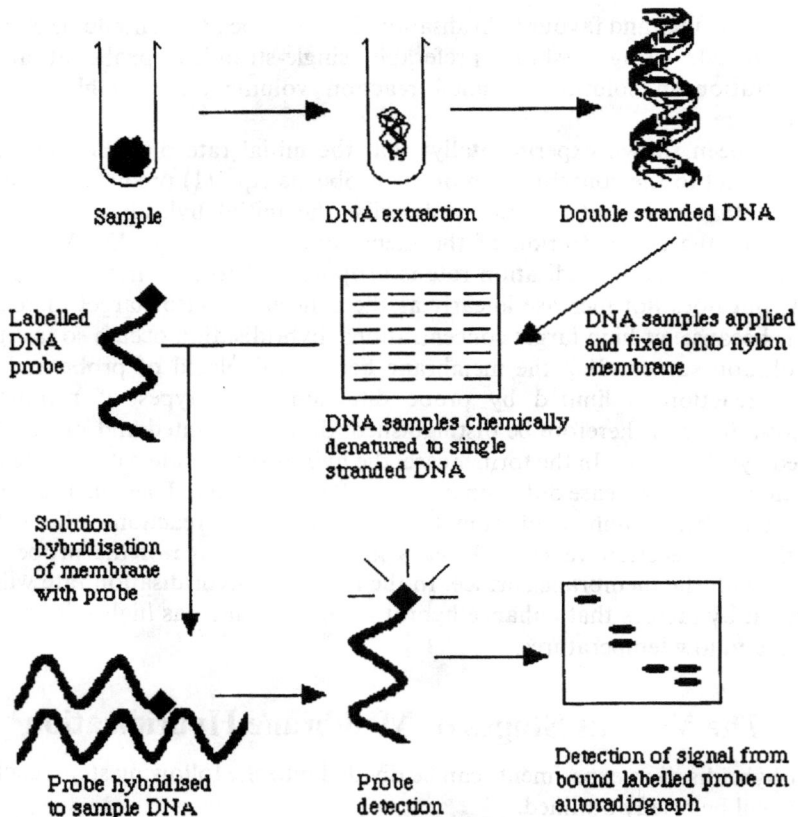

Figure 9.1   *Schematic diagram of a typical slot blot procedure*
(adapted from Hunt[2])

9.2.2   Nucleic Acid Gel Electrophoresis

Some membrane hybridisation applications, such as RFLP analysis, require the
electrophoretic separation of DNA which has been digested with restriction
enzymes prior to immobilisation onto a membrane. Electrophoresis is not
covered in this chapter, but a description of the theory behind gel electrophoresis along with detailed practical protocols can be found in Andrews[4] and a
summary of useful electrophoresis data is provided by Patel.[5]

9.2.3   Immobilisation of DNA onto the Membrane

DNA must be transferred and fixed (immobilised) onto the membrane of choice
before hybridisation. Transfer of the nucleic acid in solution to the membrane
can be carried out by simple pipetting or by using slot and dot blot apparatus.
DNA that has been size separated by gel electrophoresis is transferred from the

gel to the membrane by capillary action or under vacuum, a technique known as Southern blotting.

Various materials are used for the immobilisation of nucleic acids for hybridisation experiments. These include nitrocellulose, nylon and activated papers, each of which is suitable for different applications (reviewed by Dyson[6]). Nitrocellulose was the first widely used solid support membrane, but its use is somewhat limited by its low tensile strength and combustibility. Nylon, in contrast, is more resilient and is generally the membrane of choice today. Gilmartin[7] provides details on the different membranes available and lists commercial suppliers.

9.2.4 Preparation and Labelling of Nucleic Acid Probes

In general, the choice of label and labelling method for a particular analysis depends mainly on the type of probe and the degree of sensitivity required. Traditionally, radioactive labelling (e.g. ^{32}P, ^{35}S) was the only means of detecting probe–target hybrids. These labels are increasingly being replaced by safer, non-radioactive alternatives, such as digoxigenin, biotin and fluorescein. There are a variety of methods available for the generation of labelled nucleic acid probes, including nick translation and random priming, which generate single-stranded DNA probes; PCR labelling, which generates double-stranded DNA probes or single-stranded DNA probes (asymmetric PCR); and 3'-end labelling or tailing of oligonucleotide probes. RNA probes are commonly synthesised and labelled by *in vitro* transcription of DNA with RNA polymerase.

A comprehensive review of reporter moieties and probe labelling is beyond the scope of this chapter; many excellent reports covering extensive background protocols are available.[8–12]

9.2.5 Hybridisation

The hybridisation procedure can be divided into three distinct stages: pre-hybridisation, hybridisation and post-hybridisation washing. Prior to hybridisation, membranes are routinely incubated in a pre-hybridisation solution, which contains components such as Denhardt's solution and salmon sperm DNA, which are designed to 'block' non-specific nucleic acid binding sites on the membrane. Many different pre-hybridisation solutions have been used but a 'typical' solution might contain 6 × SSC, 5 × Denhardt's solution, 0.5% SDS and 100 $\mu g/ml$ denatured salmon sperm DNA.[7] Failure to perform this step results in high background signals arising from non-specific probe–membrane interactions. The pre-hybridisation solution is subsequently replaced by the hybridisation solution, and incubation is continued at an empirically determined optimal temperature. In general, the composition of the hybridisation and pre-hybridisation solutions are identical, with the exception that the hybridisation solution contains the labelled probe. Non-radioactive detection systems, such as the DIG (digoxigenin) system from Boehringer Mannheim, often use specific hybridisation solutions and details of these should be obtained from protocols supplied by the manufacturer. Both pre-hybridisation and hybridisation can be performed either in sealed plastic bags placed in water baths or in hybridisation tubes contained in specifically designed incubators (hybridisation ovens).

Following hybridisation, the membranes are washed to remove unhybridised probe and dissociate unstable non-specific probe–DNA hybrids which may have formed during hybridisation. The conditions and stringency of the post-hybridisation washes are generally determined empirically, and can be varied to suit the required specificity of a particular experiment. Reports covering hybridisation protocols and practical advice are available.[3,6,7]

9.2.6 Detection

Following hybridisation and washing, the bound probe–target DNA hybrids are detected. The detection method chosen is dictated by the nature of the reporter group. Radioactive labels are directly detected following exposure to autoradiographic film, whereas for non-radioactive systems the immobilised probe–DNA complexes are generally detected colorimetrically or chemiluminescently, following incubation with enzyme conjugated antibodies, specific for the reporter labels incorporated into the probes. Commonly used enzymes include alkaline phosphatase, which catalyses NBT/BCIP (nitroblue tetrazolium/5-bromo-4-chloro-3-indoyl phosphate) for colorimetric detection and CSPD® or CDP-Star™ (Boehringer Mannheim) to produce chemiluminescent signals, and horse radish peroxidase, which catalyses the oxidation of luminol in combination with hydrogen peroxide, generating a chemiluminescent signal. Alternatively, these enzymes can be directly conjugated onto the probe. Colorimetric signals are recorded directly on the hybridisation membrane, whereas chemiluminescent signals are detected by autoradiography. New products are being constantly developed and precise experimental protocols are best found in manufacturer's instructions. A more detailed description of the many detection procedures can be found in some of the specific texts available.[9,10,13]

9.3 Validation of Membrane Hybridisation
9.3.1 Introduction

The previous sections have described the basic theory behind membrane hybridisation. This theory can act as a guide to the experimental issues which may affect the result of a hybridisation experiment. However, membrane hybridisation is a complex system and it should be emphasised that it is impossible to predict experimental behaviour completely. In practical terms, precise experimental conditions are best determined by trial and error, using the underlying theory as a troubleshooting guide. The complex nature of hybridisations, and the multiple experimental variables that can affect results, make it challenging to ensure that valid results are reproducibly obtained. This section will briefly examine the practical stages of DNA hybridisation experiments and attempt to identify and highlight experimental areas which are important in ensuring valid results.

Great care should be taken throughout the hybridisation procedure to prevent damage to the membrane, for example with pipette tips when loading samples or when handling the membranes with tweezers. Damage tends to lead to high non-specific background signals. Similarly, allowing the membrane to dry out between hybridisation steps also results in high background signals and should be avoided. It is also recommended to wear powder-free gloves at all times when handling membranes to avoid fingerprints.

9.3.2 DNA Extraction

The nucleic acid obtained for hybridisation is a function of the sample material and the method of DNA extraction employed. The quality of DNA suited to hybridisation techniques may vary, depending on the methodology employed. For restriction digestion-based analyses such as RFLP, DNA quality is of prime importance. Such applications necessitate high molecular weight sample DNA of high purity, as both the loss of restriction sites associated with degraded DNA and the presence of enzyme inhibitors may lead to incomplete digestion and equivocal results. If restriction digestion is not required (for example, for dot/slot blot and colony hybridisations), DNA quality is of lesser importance and a degree of sample DNA degradation can be tolerated. In fact, DNA of sufficient quantity and quality for colony hybridisation assays can be generated by the crude *in situ* alkaline lysis of bacterial cells directly on a membrane.

The quality of the sample nucleic acid is also of particular concern in quantitative hybridisation analyses, and in particular for the quantification of suspected adulterants in processed food samples.[2] At LGC, such analyses are routinely performed by comparing the post-hybridisation signals generated following species-specific probing of known amounts (as determined by UV spectroscopy) of both sample and control DNAs (of varying concentrations), immobilised on a membrane surface. When the sample under analysis is highly degraded, the amount of DNA present is routinely overestimated by UV spectroscopy (Chapter 4). This can lead to the analyst inadvertently loading less material than anticipated onto the membrane, resulting in an underestimation of the amount of adulterant present. This observation is further illustrated in Figure 9.2, where equivalent spectrophotometrically determined amounts of DNA extracted from durum wheat and 100% durum wheat pasta were applied to a membrane and hybridised with a wheat-specific probe.

The important point is that, as far as possible, control samples should be of the same level of integrity as the sample being analysed, so that truly comparative data can be generated. Validation of experimental procedures involving sample collection and storage, and the factors affecting DNA extraction, are discussed in Chapter 3 and will not be covered here; quantification of total DNA is covered in Chapter 4.

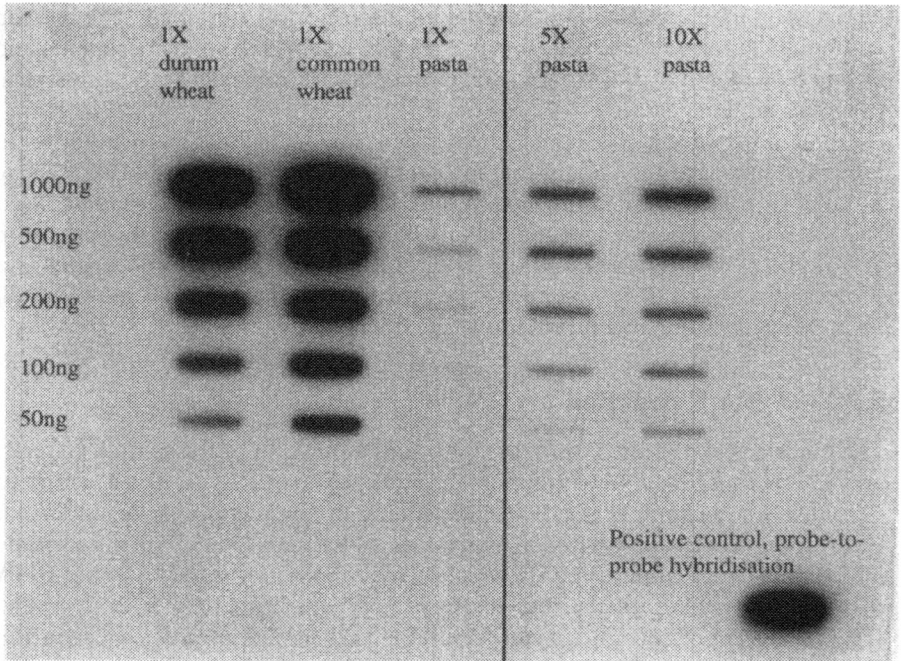

Figure 9.2 *Slot blot hybridisation of wheat and pasta DNA with a universal wheat probe. The concentration of DNA obtained from durum wheat, common wheat and pasta (made with 100% wheat) was determined using a spectrophotometer. Supposed equivalent amounts of DNA were placed on the membrane as shown in the first 3 lanes. The post-hybridisation signal of the pasta DNA is significantly less than the spectrophotometric measurements obtained. Subsequent visualisation of the pasta DNA, after agarose gel electrophoresis, revealed it was in a highly degraded state*

9.3.3 Immobilisation of Nucleic Acid to Membrane

Many factors affect the immobilisation of nucleic acid onto membranes, including the membrane type, the nature of the DNA and the conditions used for immobilisation.[6] Practical aspects of Southern blotting genomic DNA for DNA profiling are discussed by Thomson.[14]

9.3.3.1 Membrane Type

There are currently a variety of different membrane types from different manufacturers available to the analyst, including both nitrocellulose- and nylon-based membranes. Each has different requirements for optimal nucleic acid immobilisation and displays different binding capacities.[7] As a consequence, protocols developed with one type of membrane are not necessarily transferable to another. For example, nucleic acids bind more efficiently to nitrocellulose membranes under high salt conditions, while for some nylon membranes, optimal nucleic acid binding is achieved at lower salt conditions. It is therefore important to follow the manufacturer's recommended instructions for optimal nucleic acid transfer and binding efficiency.

9.3.3.2 Transfer of Nucleic Acid to Membrane

Nucleic acids must be transferred to the membrane surface prior to immobilisation. For the most simplest of applications, a nucleic acid sample can be applied directly onto a hybridisation membrane either by pipetting a small aliquot (1–2 μl) onto the membrane surface or, when larger volumes or more samples are required, using specialised vacuum-assisted dot/slot blot apparatus. The more complex Southern blotting technique involves the transfer of DNA onto the membrane following electrophoretic fractionation of restriction fragments. This was traditionally preformed using capillary blotting, but, increasingly, vacuum generating devices are being used to speed up the transfer process.

9.3.3.3 Method of Immobilisation

One of the critical steps in solid-phase hybridisation assays is the immobilisation of nucleic acids on the solid support. Traditionally, membranes were baked at 80 °C (under vacuum for nitrocellulose and under normal air conditions for nylon) to complete nucleic acid immobilisation. However, this mode of attachment is non-covalent and, as a result, DNA can be lost from the membrane surface by diffusion, particularly under conditions of extended hybridisation times, elevated temperatures and repeated stringent washes. This is most evident with 'probe-sized' DNA fragments, which may dissociate from the membrane following hybridisation with the probe.[3] These events may lead to problems interpreting quantitative hybridisations. The problem of DNA loss can be overcome by the use of UV crosslinking. This process results in the formation of covalent linkages between the DNA and the membrane surface, and ensures all fragments are bound irrespective of size. The covalent binding, coupled with the durability of nylon, allows hybridised probe to be removed by stringent washing, permitting multiple re-hybridisations of the sample DNA, which is retained on the membrane. UV crosslinking should not be used with nitrocellulose membranes owing to the risk of fire. Alternatively, positively charged nylon membranes can be used, which bind DNA covalently following alkaline treatment, negating the need for baking or UV treatment.

Immobilisation of the nucleic acid sample onto the membrane can greatly affect the eventual sensitivity of the analysis. If fixation is inadequate, target nucleic acid can be lost from the membrane during hybridisation, stringent washing or re-probing, leading to decreased sensitivity, whereas a prolonged fixation may result in damage to the nucleic acid. To achieve maximum sensitivity, it is therefore essential that the optimal immobilisation conditions are consistently used.

9.3.3.4   Size and Nature of the Nucleic Acid

The efficiency of nucleic acid transfer and immobilisation is affected by its size, and this can have important consequences on subsequent quantitative analyses. Efficient transfer of high molecular weight DNA from gels to membrane supports is difficult to achieve, and partial fragmentation of large DNA molecules by HCl treatment (depurination) is generally required prior to transfer to ensure equal transfer efficiencies of all DNA molecules. This step is of particular importance for the quantitative transfer of DNA fragments > 5–6 kbp. However, incubation times must be adhered to closely, as excessive depurination will generate very short fragments which may be poorly retained by the membrane. Small DNA fragments (200 bp or less) do not transfer efficiently from agarose to nitrocellulose, and nylon membranes should be used to prevent under-representation of such samples. High molecular weight DNA does not tend to bind well to membranes and binds more efficiently when fragmented with restriction enzymes (as in Southern blotting) or sheared into smaller sections (for example, by sonication). Highly degraded DNA may also be problematic as its small size may allow it to pass through certain membranes without binding. In general, nylon membranes bind small nucleic acids more efficiently than nitrocellulose, although care should be taken with material below 100 bp in length. Again, because the fundamentals of membrane hybridisation are unclear, it is difficult to be precise about the affects of DNA size differences on experimental performance.

DNA is routinely denatured prior to immobilisation, as membranes tend to bind single-stranded DNA more efficiently. This also ensures that the DNA is immobilised in a single-stranded form ready for hybridisation. Nucleic acid denaturation can be performed either before (dot/slot blot), during (Southern blot) or after (dot/slot blot) samples are applied to the membrane, and is generally achieved by alkaline treatment.

9.3.4   Nucleic Acid Probes

In order to achieve valid results and maintain consistency between separate experiments, the probe used for related applications and successive repeats of a particular hybridisation should ideally be identical in composition and present at the same concentration. The following points may therefore need to be considered.

9.3.4.1   Probe Production and Labelling

There are many potential sources of variation in probe labelling, which must be carefully monitored in order to generate reproducible and valid analytical data.

Methods of probe labelling such as nick translation and random priming generate a random population of truncated probe molecules, which are unlikely to be consistent between different labelling reactions. This heterogeneity in probe population is a potential problem in ensuring consistent results. In addition, owing to the nature of the synthesis and labelling process, a fraction of the probes generated by random priming will share partial complementarity and may therefore form concatamers in the hybridisation solution. This can have complicated effects if quantification is required (refer to Section 9.3.5.4).

The efficiency of probe labelling is likely to vary between successive labelling reactions, and must be determined prior to hybridisation (see Section 9.3.4.2). In general, the higher the purity of the DNA template, the better the labelling efficiency; therefore it is important to ensure that the probe or template DNA is as free from contaminants as possible. It is also important to ensure the complete removal of any unincorporated labelled nucleotides from the labelled probe solution, as these can contribute significantly to background 'noise'. Other reaction components which may interfere with hybridisation, for example salt, must also be removed prior to use. Removal of these potential contaminants is generally achieved by gel filtration or ethanol precipitation.

Furthermore, PCR-generated probes have the potential to contain labelled non-specific amplification products which could cross react with non-target membrane-bound DNA. These non-specific PCR products can be largely eliminated by careful optimisation of the PCR parameters (Chapter 5).

Probe labelling time is an additional criterion that can have an overall effect on assay reproducibility. For enzyme-mediated probe labelling in particular, it is important that optimum timing is determined and strictly adhered to, as deviation may result in significant variation in the total amount of label incorporated into the probe molecule.

9.3.4.2 *Probe Concentration*

In general, hybridisation rate and, within narrow limits, sensitivity increase with probe concentration. For double-stranded probes, however, excessively high concentrations of probe in solution will favour probe reassociation over hybridisation to membrane-bound DNA. For single-stranded probes, there is no reassociation in solution and the rate of hybridisation to membrane-bound DNA will therefore increase with increasing probe concentration. There are, however, limits to the extent to which probe concentration can be increased before non-specific binding to the membrane occurs. This problem is thought to be due to the nature of the probe reporter moiety and the non-specific binding properties of the membranes rather than any intrinsic properties of the probe itself.

In order to avoid the problem of background staining, the optimal concentration limits for all probes should be determined empirically. Boehringer Mannheim[15] recommend carrying out a 'mock hybridisation' (hybridisation procedure performed on membranes without DNA) prior to hybridisation experiments to evaluate the optimal probe concentration, i.e. the highest probe concentration that results in the most acceptable background. For membrane hybridisations using DIG-labelled probes, Boehringer Mannheim[15] recommend concentrations of 5–25 ng/ml for DNA probes, 100 ng/ml for RNA probes and between 0.1 and 10 pmol/ml for oligonucleotides.

In practice, it can be difficult to determine the exact concentration of labelled probe added to the hybridisation, and it is common to add a set percentage volume of the completed labelling reaction to the hybridisation solution without taking into account labelling efficiency. In the majority of cases, labelling efficiency is rarely 100%, and is likely to vary between successive labelling reactions. For these reasons, it is extremely important to check the efficiency of all labelling reactions if valid and consistent results are to be generated. For isotopic procedures, incorporation of radiolabelled nucleotides is generally determined by a direct measurement of the probe's specific radioactivity. For non-isotopic labelling procedures, it is not usually practical to measure the incorporation of modified nucleotides or reporter groups directly, and the degree of labelling can be measured indirectly, by comparing signals generated from dilutions of control DNA and experimentally labelled probe. Ideally, the control and probe DNA should be similar in composition and size, but this is not always the case. Even if this is carried out, it is possible that each labelling reaction contains variable amounts of unlabelled probe DNA that can compete with labelled probe for the target DNA and complicate the interpretation of results. The consequences of not controlling this variable are potentially serious. For example, in quantitative approaches, where measurement is based on the signal generated following a certain hybridisation time, variation in the concentration of labelled to unlabelled probe can lead to variations in the response obtained.

The regular labelling of small quantities of probes is generally associated with non-reproducible data, as the labelling efficiency, and hence consistency, is likely to vary between different 'batches'. In order to maintain consistency of labelling and hence generate comparable and reproducible results, it is therefore recommended to synthesise and label large quantities of probes at one time. This approach, however, is not applicable to radioactively labelled probes, whose routine use is limited by a short shelf-life, for example 14.2 days for ^{32}P. Reporter-labelled probes, however, have a much longer shelf-life, and can be stored for up to a year at $-20\,^{\circ}$C without loss of activity. For chemically labelled probes, the consistency of labelling is more reproducible and controllable than enzyme-mediated methods, and the procedures used are more suitable for yielding large quantities of labelled probes. In addition, chemical methods generally employ post-labelling purification methods, e.g. reverse phase HPLC, to remove unlabelled DNA from the labelled probe solution. However, as enzymatic-labelling techniques generally result in a greater degree reporter molecule incorporation over chemical labelling, overall sensitivity of detection is generally greater for enzymatically labelled probes.

9.3.5 Hybridisation and Washing

To reduce experimental variation and ensure repeatable and valid results it is important to minimise the variables that can affect nucleic acid hybridisation. Effective hybridisation is a function of the hybridisation rate and the stability of the nucleic acid hybrids once they are formed (defined by their T_m), and the understanding these will enable the analyst to derive hybridisation conditions which should yield the optimal signal-to-noise ratio.

A variety of parameters, including ionic strength, hybrid base composition and concentration of formamide in the hybridisation solution, can affect the hybrid T_m and hybridisation rate and thus have an overall consequence on the hybridisation reaction itself. These are discussed below.

9.3.5.1 Temperature and Ionic Strength: Effects on Hybridisation Rate

A plot of hybridisation rate against temperature typically produces a bell-shaped curve (Figure 9.3). At lower temperatures, DNA–DNA hybridisation occurs extremely slowly, but increases dramatically as the temperature is increased until a maximal rate of hybridisation is reached, which is 20–25 °C below the T_m of the hybrids. At higher temperatures the hybridisation rate decreases as the probe–target duplexes are unstable and tend to dissociate. Therefore, DNA–DNA hybridisations should ideally be performed at a

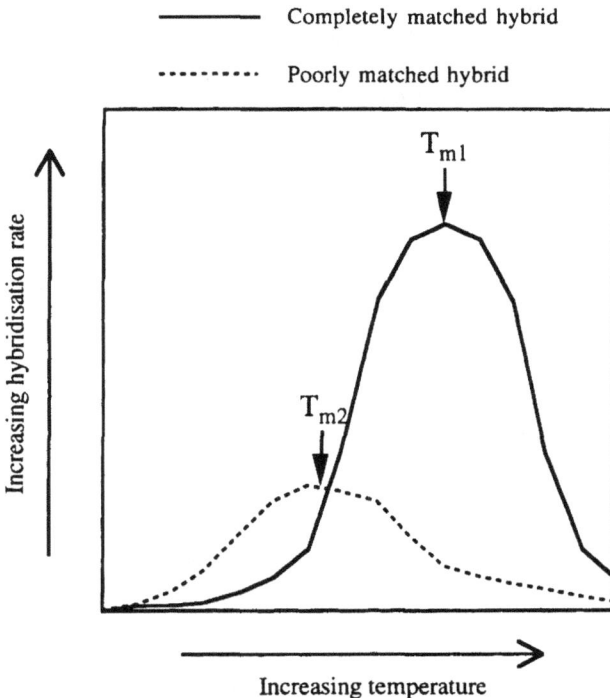

Figure 9.3 *Theoretical rate of hybridisation as a function of temperature.* T_{m1} *is the melting temperature of a completely matched hybrid and* T_{m2} *is the melting temperature of a mis-matched hybrid*

temperature that is 20–25 °C below the duplex T_m. In practice, hybridisation is normally carried out at 68 °C in aqueous solution and 42 °C for solutions containing 50% formamide (for well matched probes over 100 base pairs in length). For oligonucleotides, hybridisation is usually performed at a temperature 5 °C below the duplex T_m.

The ionic strength of the hybridisation and washing solutions, normally measured as the concentration of monovalent cation (sodium), plays a major role in determining the rate and stringency of hybridisation. When the hybridisation solution contains little or no salt, nucleic acids hybridise very slowly, but as the ionic strength increases, the rate of hybridisation increases. The maximum hybridisation rate generally occurs with sodium concentrations of approximately 1.5 M and continues to rise, at a much slower rate, up to around 4 M.

9.3.5.2 Temperature and Ionic Strength: Effects on Hybridisation Stringency

The purpose of many hybridisation experiments, particularly in clinical or diagnostic applications, is the unequivocal identification of a particular nucleic acid sequence. The potential problem of non-specific hybridisation with non-target nucleic acids is of obvious concern. It is often the case that the sample nucleic acid immobilised on the membrane is a mixture of sequences of which only a small percentage is the target. These will range from sequences completely unrelated to the target sequence to those sharing a high degree of similarity.

In order to minimise the hybridisation of probe to related but non-identical sequences, hybridisation and post-hybridisation washes must be performed under the most stringent conditions possible. Stringency is determined by a combination of temperature, salt and/or formamide concentration, and can be controlled either during hybridisation or in the post-hybridisation washes.

- *Temperature* — The 'bell-shaped' relationship between the rate of hybridisation and the temperature of incubation discussed in Section 9.3.5.1 is still followed for hybrids which are not perfectly matched, but the rate constant is lower and the curve is displaced towards lower temperatures, i.e. the T_m reaches its optimum at a lower temperature (Figure 9.3). From this it is apparent that to distinguish closely related sequences (those with a higher chance of cross-hybridisation) it is necessary to hybridise at high temperatures where the ratio of mis-match hybridisation to homologous hybridisation is low. In contrast, if the purpose of the experiment is to identify related nucleic acid sequences (such as screening a library), then hybridisation should be carried out under less stringent conditions, such as reduced temperatures.
- *Ionic strength* — High salt concentrations stabilise mis-matched duplexes, so to detect distantly related but non-identical nucleic acid sequences

(cross-reacting species), the salt concentration of the hybridisation and washing solutions must be kept fairly high. In contrast, to distinguish closely related nucleic acid sequences, hybridisation and washing must be performed under low salt conditions (high stringency), where mismatched duplexes are not tolerated. In practice, high stringency washes are generally carried out using 0.5–2 × SSC, and low stringency, using 2–6 × SSC.

In practice, stringency is more usually determined by the post-hybridisation washing conditions than by the hybridisation temperature itself, and can easily be altered to suit the requirements of a particular analysis.

9.3.5.3 Temperature Control

As evident from the above paragraphs, maintaining the optimum temperature throughout hybridisation and subsequent washing steps is critical to ensuring experimental specificity and repeatability. This is often neglected during experiments. For example, when using glass hybridisation tubes, it can take a considerable time for the temperature of the solution in the tubes to reach the set oven temperature. Similarly, it is common for wash solutions in sandwich boxes (a common way to carry out washes) to be at a lower temperature than the water bath in which they are incubated. This may have important consequences in particular situations where the stringency of a particular wash is critical in achieving the required discrimination.

Overcoming these problems relies on accurate monitoring of reaction temperatures. Hybridisation ovens and water baths should be routinely calibrated, and temperatures should be regularly checked with a thermometer to ensure they are correct. This should include temperatures of the hybridisation solution inside glass hybridisation tubes, and wash solutions inside sandwich boxes, to ensure the solution temperature is the same as the set instrument temperature. Instruments, containers and solutions should be given ample time to reach experimental temperatures prior to use and care should be taken when moving solutions to reaction vessels of a different temperature in case of temperature change due to heat transfer to and from the reaction solutions. Care should be taken to ensure that, at the high temperatures required for many hybridisations and subsequent washing steps, evaporation of water does not occur, leading to an increase in salt concentration. This is most likely to occur in sandwich boxes contained in water baths where the top of the sandwich box is cooler than the incubating water. This encourages condensation of the evaporated liquid on the sides of the box. The problem is accentuated by the use of small volumes. Measures to prevent evaporation should be adopted, such as conducting the reactions in sealed vessels with no air space (plastic bags) and using systems where the reaction vessel is uniformly the same temperature as the hybridisation solution (glass tubes in hybridisation ovens).

9.3.5.4   Other Components of Hybridisation Solutions

Formamide, a helix destabiliser, decreases the T_m of nucleic acid hybrids. Its use allows the hybridisation temperature to be reduced, typically to 30–42 °C with 30–50% formamide, without affecting hybridisation stringency. This has important practical advantages: the probe and membrane bound nucleic acid (particularly RNAs) are more stable at lower temperatures, making it less likely that degradation will occur during hybridisation. Excessively high formamide concentrations (80%), however, reduce the hybridisation rate significantly, as DNA–DNA hybrids are not stable under these conditions.

The hybridisation rate of long probes (> 250 bp) can be accelerated by the inclusion of inert polymers, e.g. dextran sulfate and poly(ethylene glycol) in the hybridisation solution. Accelerating rates are relatively low with single-stranded probes (three-fold), but increases up to 100-fold are observed with double-stranded probes. The underlying mechanism of this effect is thought to be due to two factors: (i) the exclusion of probe from the volume of hybridisation solution occupied by the polymer, effectively increasing the concentration of probe, and (ii) the formation of concatenates, networks formed when probe fragments hybridise with each other in overlapping complementary regions.[16] The latter may prove problematic in quantitative studies, as the intensity of signals generated following hybridisation of these probe networks to membrane-bound DNA will depend upon the number of labelled probes attached to each target. The variability in the size of probes generated by nick translation and random priming may also lead to inconsistencies in the dextran sulfate effect, as smaller probes can only overlap over short regions and will therefore be unable to maintain stable concatenate structures. Optimal concentrations of hybridisation accelerators may vary between individual hybridisations, and should be determined empirically.

Increasing viscosity of the hybridisation solution leads to a reduction of the hybridisation rate. The mechanism is complex and not well understood. Viscosity is increased by components such as glycerol and sucrose. Although components such as dextran sulfate and poly(ethylene glycol) lead to increased viscosity, the decrease in hybridisation rate from the viscosity change is more than offset by the increase in rate caused by these accelerators.

The majority of hybridisation experiments are carried out in the pH range 5–9 and, at normal salt concentrations, the impact of pH on the rate of hybridisation is very slight.

9.3.5.5   Time Period of Hybridisation

Incubation time for hybridisation is an important variable in hybridisation assays, and is an important consideration in the ability to distinguish between related sequences. Intuitively, it might seem that if a hybridisation is allowed to continue for an extended period of time, the reaction is more likely to be specific, as the probe has more time to find its correct target sequence. This, however, is not the case. With both single- and double-stranded probes in excess (a common situation in standard Southern, Northern and dot/slot blots), prolonged hybridisation does not increase the extent of hybridisation as more and more probe reassociates or binds non-specifically to the membrane, at the expense of hybridisation to the true target. In addition, at the high temperatures involved, non-covalently bound sequences may leach off the membrane and the probe may degrade. In practical terms, hybridisation times should therefore be kept as short as necessary to obtain sufficient specific signal on detection. In contrast, when the membrane-bound sequence is in excess (for example, when probing plasmid DNA), the ability of the probe to discriminate non-target from target sequences remains constant and is not affected by the time of incubation. The majority of the probe is hybridised to the target sequence (low ratio of cross-hybridisation to self-hybridisation) and only a small percentage of the probe is reassociated because the membrane-bound sequences are continually competing for the same limiting probe. In summary, to distinguish between cross-hybridising species in practice, it is best to use short incubation times regardless of whether the probe or the filter-bound sequence is in excess. If this does not generate enough signal to be detected, the amount of membrane-bound target should be increased as opposed to hybridising for longer times.

Hybridisation time is probably the most poorly regulated of experimental variables in hybridisation experiments. The statement 'overnight incubation' is widely used, which is very imprecise. Experiments hybridised for variable times are often thought of as equivalent with little thought for the consequences of the experimental variation. To improve the situation, more stringent adherence to set hybridisation times needs to be adopted.

9.3.5.6   Re-use of Probe and Membranes

Normally, only a very small percentage of the probe is used up during hybridisation, and it is common for analysts to decant the probe/hybridisation solution into a separate vessel after hybridisation, and store it at $-20\,^{\circ}\text{C}$ for re-use in subsequent analyses. However, in terms of generating consistent and reproducible data, it is important to determine the optimal number of times a single hybridisation/probe solution can be used before the probe becomes too dilute, degraded or, in the case of radiolabelled probes, loses specific activity, and the overall sensitivity of detection is impaired.

Re-probing of membranes with different probes is also common, particularly in RFLP, where a series of probes are routinely used to generate informative data from a single Southern blot. Before re-probing, it is necessary to strip off all the old probe from the membrane-bound DNA fragments. Incomplete removal of probe can lead to misleading results, especially if the old probe is still detectable after a second round of hybridisation, or if it prevents the annealing of the new probe to its target sequence. It is therefore critical to determine optimal stripping procedures for particular probes, and to monitor the effectiveness of probe removal prior to re-hybridisation with a different probe. In addition, it should be noted that the prolonged use of membranes leads to a gradual reduction in sensitivity, and only data generated prior to this point are comparable.

9.3.6 Detection

Autoradiography is the most common method used to detect probes following hybridisation and is applicable to both radioactive and chemiluminescent labelling methods, providing a permanent record of the results. In addition, by varying the exposure time, it is possible to generate multiple hard copies of different signal intensity. This may be useful when bands of significantly differing strengths are to be detected on the same membrane. The time required to expose X-ray film to visualise light emission is directly dependent on both the type and amount of label present in the sample. Therefore, when comparing individual hybridisation assays, it is important to hybridise each membrane with identical amounts of labelled probe, preferably from the same batch, and to monitor light production over identical time periods.

A significant limitation of autoradiography is based on the fact that X-ray films have a threshold for both response and sensitivity and a relatively low saturation level, and therefore display a non-linear response to hybridisation signal. This makes comparison and quantitation of results extremely difficult. One option is to pre-flash film, which leads to an improved linear response.[17]

An alternative to autoradiography is the use of molecular imagers (such as the Biorad GS-525 and the Molecular Dynamics 'Storm' system), which use a phosphor screen to record the hybridisation signal. The screen is then scanned with a pulsed laser scanner in order to collect the data for analysis and display. The most important advantages of this technology over the use of X-ray films are enhanced sensitivity and a linear response of the system over a very wide range of signal intensity. This means, for example, that if a hybridisation signal generated using molecular imaging technology for a particular sample is 100 times more intense than another, you can say with much greater confidence that there is indeed 100 times the amount of target sequence present. This has obvious implications for accurate quantitative hybridisation assays, although the application of this system is somewhat limited by its high cost.

Unlike radioactive- and chemiluminescent-based detection procedures, colorimetric detection generates a visible product on the membrane surface. As detection occurs directly on the membrane, it is extremely difficult to generate a permanent record of signal intensities over varying colorimetric reaction times. This can be achieved by photographing the membrane at various stages of the substrate reaction, but this approach is generally time consuming. In addition, the coloured precipitates are often very difficult to remove, making re-probing extremely difficult.

For both colorimetric and chemiluminescent detection procedures, it is recommended to include a 'test strip' in the detection assay. This is a small strip of membrane, onto which known amounts of appropriately labelled DNA have been immobilised, which is useful in monitoring the success or failure of the immuno-enzymatic reaction and therefore potentially eliminating false negative results caused by a faulty or badly executed detection procedure.

9.4 Summary

The ubiquitous use of membrane hybridisation in molecular biology tends to play down the complex nature of the technique. In reality, obtaining accurate and reproducible results is dependent on the successful completion of multiple experimental steps and a clear understanding of how the process is affected by experimental parameters. This chapter has highlighted some of the key issues that should be considered and will hopefully prove to be useful in the design and validation of membrane hybridisation-based analytical tests.

9.5 References

1. Jeffreys, A. J., Wilson, V. and Thein, S. L. 1985. Individual-specific 'fingerprints' of human DNA. *Nature* **316**: 76–79.
2. Hunt, D. J., Parkes, H. C. and Lumley, I. D. 1997. Identification of the species of origin of raw and cooked meat products using oligonucleotide probes. *Food Chem.* **60**: 437–442.
3. Anderson, M. L. M. and Young, B. D. 1985. Quantitative filter hybridisation. In: Nucleic Acid Hybridisation. A Practical Approach (eds. Hames, B. D. and Higgins, S. J.), chap. 4. IRL Press, Oxford.
4. Andrews, A. T. 1991. Electrophoresis of nucleic acids. In: Essential Molecular Biology. A Practical Approach (ed. Brown, T. A.), vol. 1, chap. 5. IRL Press, Oxford.
5. Patel, D. 1994. Gel Electrophoresis. Essential Data Series (eds. Rickwood, D. and Holmes, B. D.). Wiley, Chichester.
6. Dyson, N. J. 1991. Immobilisation of nucleic acids and hybridisation analysis. In: Essential Molecular Biology. A Practical Approach (ed. Brown, T. A.), vol. 2, chap. 5. IRL Press, Oxford.
7. Gilmartin, P. M. 1996. Nucleic Acid Hybridisation. Essential Data Series (eds. Rickwood, D. and Holmes, B. D.). Wiley, Chichester.
8. Arrand, J. E. 1985. Preparation of nucleic acid probes. In: Nucleic Acid Hybridisation. A Practical Approach (eds. Hames, B. D. and Higgins, S. J.), chap. 2. IRL Press, Oxford.
9. Mundy, C. R., Cunningham, M. W. and Read, C. A. 1991. Nucleic acid labelling and detection. In: Essential Molecular Biology. A Practical Approach (ed. Brown, T. A.), vol. 2, chap. 4. IRL Press, Oxford.
10. Kricka, L. J. (ed.). 1992. Nonisotopic DNA Probe Techniques. Academic Press, San Diego, CA.
11. Keller, G. H. and Manak, M. M. (eds.). 1993. DNA Probes, 2nd edn. Stockton Press, New York.
12. Manak, M. M. 1993. Radioactive Labelling Procedures. In: DNA Probes (eds. Keller, G. H. and Manak, M. M.), 2nd edn., sect. 4. Stockton Press, New York.
13. Manak, M. M. 1993. Hybridization Formats and Detection Procedures. In: DNA Probes (eds. Keller, G. H. and Manak, M. M.), 2nd edn., sect. 6. Stockton Press, New York.
14. Thomson, J. 1998. Southern blotting of genomic DNA for DNA profiling. In: Forensic DNA Profiling Protocols (eds. Lincoln, P. J. and Thomson, J.), pp. 49–56. Humana Press, Totawa, NJ.

15. Boehringer Mannheim 1993. The DIG System User's Guide for Filter Hybridisation. Boehringer Mannheim GmbH, Biochemica, Germany.
16. Wahl, G. M., Stern, M. and Stark, G. R. 1979. Efficient transfer of large DNA fragments from agarose gels to diazobenzyloxymethyl-paper and rapid hybridisation by using dextran sulfate. *Proc. Natl. Acad. Sci. USA* **76**: 3683–3687.
17. Laskey, R. A. and Mills, A. D. 1975. Quantitative film detection of ^{3}H and ^{14}C in polyacrylamide gels by fluorography. *Eur. J. Biochem.* **56**: 335–341.

CHAPTER 10

Automated Fluorescent DNA Cycle Sequencing

N. J. OLDROYD AND I. L. COMLEY,
PE Biosystems (a division of Perkin-Elmer)

10.1 Introduction

Traditionally, DNA sequencing has involved the use of manual protocols which, though relatively inexpensive to perform, are labour intensive. The demands of current projects require large amounts of accurate and reproducible sequence data from increasing numbers and varieties of templates. As a result, manual methods can be limiting, both in terms of the time taken and the amount of data which can be obtained from each gel. In addition, the interpretation of manual sequencing data is usually carried out by eye which can impact upon the consistency and reproducibility of results. Such manual protocols, though still in existence, tend to represent the history rather than the future of DNA sequencing.

Efforts to improve DNA sequencing have not been confined to apparatus. Specific DNA polymerase enzymes have been developed to provide more robust sequencing protocols. This began with heat labile DNA polymerases which gave uniform, easy to interpret bands. These enzymes, however, have their limitations, particularly when confronted with secondary structures in the template. If the enzyme is not able to process through the secondary structure it will drop off the template, producing erroneous bands. As a result, these enzymes have largely been replaced by the use of thermostable polymerases[1,2] which function at much higher temperatures; under such conditions the template is much less likely to form secondary structures. The stability of these enzymes confers the further advantage that the reaction product can be melted away from the template, making it available for further reactions. The high temperatures and the repeated cycling of this sequencing approach means that these new enzymes provide a robust protocol for sequencing very low amounts of DNA from a variety of sources.

The advent of fluorescent sequencing techniques represents the most significant advancement in DNA sequencing to date. Fluorescent DNA sequencing allows the automation of data collection and analysis. As a result, fluorescent sequencing protocols have, over recent years, become the accepted means of obtaining large amounts of accurate and informative sequence data for a variety of applications. Not least amongst these is the task of sequencing the entire genomes of several organisms, including our own. It is for these reasons that this chapter will deal with the generic validation issues arising during the development of a fluorescent DNA sequencing protocol.

Fluorescent DNA sequencing produces data that are consistent, reproducible and reliable and are sufficiently robust to cope with considerable variations in template quantity and quality. However, it must be remembered that the system collects the data in real time; therefore, should the fluorescent signal generated be too weak, too strong or not of sufficient quality, the corresponding data quality will suffer. In this manner it differs from manual sequencing, which allows the user to re-expose autoradiograms to generate a stronger or weaker signal. Much of the process has been optimised by instrument design and sequencing kit formulation. This ensures the robustness of the system, but the user still plays an important role in data quality control. There are a number of factors that can affect the quality of sequence data generated, and it is these factors which we will consider in this chapter. Many of the general validation considerations relevant to fluorescent DNA sequencing, such as the use of appropriate controls, the history and composition of the sample, the time and resources available for analysis and equipment maintenance, are discussed in previous chapters. Specifically, we will discuss those factors that directly affect the quality and validity of fluorescent sequencing data.

10.2 Development of a Fluorescent DNA Sequencing Protocol

There are many factors that could conceivably affect the quality of fluorescent sequencing data; however, the major considerations can be classed under the following headings.

- *Template preparation* — There are a large number of factors in the preparation of the template that can affect the quality of the sequence data. Of particular concern is the possible contamination of the template DNA. This contamination may be due to the source of the DNA template (for example, the contamination of a plasmid preparation with genomic DNA) or may have been introduced during the preparation of the template (for example, contamination of a DNA preparation by salt).
- *Set up of the sequencing reaction* — Considerations relating to the type of chemistry, sequencing primer design, input primer, template concentration and the type of thermal cycler used can all impact on result quality.
- *Variation in template sequence composition* — Templates which display extreme characteristics (for example, high GC content or long homo-

polymer repeats) may require protocol modifications to enhance data quality.

- *Post-reaction clean-up procedures* — Carry over of large amounts of reaction constituents can reduce the quality of the data. This carry over can be controlled by following a few simple rules.
- *Reaction product detection* — Most systems rely on the use of poly-acrylamide slab gels to resolve the sequencing fragments. The software that analyses and interprets the fluorescent sequencing data requires the gels to run at certain speeds; therefore it is necessary to be able to prepare gels of a reproducible and uniform nature.
- *Data analysis and interpretation of results* — High quality data result from a process that has been optimised for each of these factors; however, there will be occasions when some of these factors lie outside the users control, for example sequence composition. A system which is optimised in all other respects can often accommodate this variation and ensure acceptable data quality. The routine use of well characterised controls allows verification of results from unknown samples.

10.2.1 Template Preparation

The quality of the template DNA can have a dramatic impact on sequence data quality. For the purposes of fluorescent sequencing, high quality template should be as free as possible of contaminants and inhibitors that could interfere with the cycle sequencing (see Chapter 6). These could have been carried over from the template preparation, such as salts, EDTA, RNA and chromosomal DNA, or include the presence of primer dimers in a PCR-generated template.

Sequencing template DNA can be inserts contained in vectors such as M13, plasmid and phage. The methods by which these can be isolated are numerous; in-house methods and commercial kits can all be used to generate template of suitable quality if correct attention is paid to the procedure. Alternatively, sequencing templates can be generated using PCR. PCR produces a template that is clean and uniform in its quality; as a result it is increasingly being viewed as a universal method of template preparation.

10.2.1.1 Plasmid DNA as a Sequencing Template

Factors influencing the validity of fluorescent DNA sequencing protocols as a result of plasmid preparation methods can be identified as listed below.

- *Selection of a suitable growth environment for template yield* — The important factor to remember when growing plasmid or phage cultures for sequencing is that only very small amounts of DNA are required and as a result it is often better to sacrifice the yield for better quality DNA. It is therefore unnecessary to use rich growth media like Superbroth; the standard Lauria broth will suffice.
- *Selection of a suitable growth environment for template quality* — Bacterial

strains vary in many respects, including expression of endonucleases and levels of glycoproteins. Such variations affect their suitability as hosts for DNA sequencing templates. For example, XL-1, DH5-alpha and HB101 host strains consistently produce good results, MV1190 tends to display variability in result quality, whereas JM101, which exhibits a high carbohydrate load, generally fails to produce plasmid template of sufficient quality for successful analysis.

- *Template purification method* — The aim of the template purification procedure is to isolate the template from the rest of the bacterial constituents. There are numerous protocols and specialised kits available to address the important aspects of the isolation. These include ensuring that no other nucleic acid is isolated other than the desired vector and that no contaminants are carried over into the sequencing reaction. A standard alkaline lysis protocol with RNAse followed by a PEG precipitation can yield excellent results without the need for costly commercial preparations. Commercial preparations can, however, provide the user with a rapid, quality controlled method of isolation. As with in-house methods, the commercial kits vary in the way in which they isolate the DNA. Ion-exchange resin-based column isolation protocols[3] selectively bind the plasmid and, provided the column is not overloaded, appears to be a robust method. In this approach the plasmid is removed from the column with a high salt concentration wash. Problems can occur if the salt is carried over into the sequencing reaction; however, ethanol precipitation is usually sufficient to desalt the template. Other commercial columns using silica to bind DNA selectively in a high chaotropic salt solution also run the risk of salt contamination, with the added possibility of silica finings being carried through with the final template. Salt will impair the action of DNA polymerase, reducing the efficiency of the sequencing reaction, whilst silica will bind to the enzyme, destroying its activity completely. To address the specific template quality issues raised by a fluorescent sequencing protocol, the ABI Prism® Miniprep kit[4] was developed to provide a fast and efficient means of producing pure plasmid.

10.2.1.2   *PCR Products as Sequencing Templates*

Using PCR products as sequencing templates enables rapid generation of sequence data from DNA or RNA, even when the sample available is in low amounts. It can also be considered as a method for preparing sequencing template from plasmids or phage. Sequencing a PCR product directly has an advantage over sequencing cloned templates from a vector. If the product is cloned prior to sequencing then the data obtained will represent the sequence from a single molecule of DNA. If, during the PCR, a random misinsertion had taken place in that molecule, it will be shown in the data obtained. It is important therefore to take a consensus of the sequences; this is achieved either by sequencing a number of the cloned products or sequencing the PCR product itself. In sequencing the product, any low level of misinsertion will contribute only at a background level and so will not be reported.

After amplification, PCR products exist in the presence of PCR primers, dNTPs, enzyme, buffer components and possibly amplification artefacts such as primer dimers and non-specific products. The influence of these potential contaminants can be reduced by ensuring that, as far as possible, the reaction components are exhausted during the reaction. This is part of normal process of PCR optimisation.[5,6] If the PCR is not optimised, the product must be purified before sequencing.

Factors influencing the validity of fluorescent DNA sequencing protocols as a result of PCR product generation can be identified under the following headings.

- *Amount of genomic DNA used in PCR* — The amount and quality of genomic DNA or other starting nucleic acid can have a substantial effect on the sensitivity and specificity of the amplification. Too little genomic template may not produce enough PCR products for fluorescent sequencing whereas too much can increase the formation of non-specific products, which could obscure the target sequence. Titration experiments can be performed to establish the optimum amount. Applications involving the amplification of rare copy targets can present certain risks. Poor fidelity of the DNA polymerase used for PCR amplification can introduce a polymorphism. If such an event occurs during the first few cycles of PCR and only limited copies of the starting gene are present, an artefactual polymorphism can be sufficiently amplified for PCR fragments containing the changed base to be detected during sequencing.
- *PCR optimisation for direct sequencing* — When performing direct sequencing, the PCR products are not purified prior to cycle sequencing. An aliquot of the PCR product is simply diluted in water to lower the concentration of residual primers and nucleotides. If the PCR is fully optimised, most of the primers and nucleotides will have been incorporated. This method can be used for both dye primer and dye terminator chemistries. To obtain the cleanest sequencing data, it may be necessary to reduce the primer and nucleotide concentrations used in the PCR amplification. This reduction is not always necessary as specific PCR reactions will exhaust reagents at different rates; therefore each reaction type will need to be optimised individually to ensure a valid result. To aid this process, PCR optimisation kits can be obtained from a number of commercial suppliers, for example the PCR Optimiser Kit from Invitrogen, the Opti-Prime PCR Optimisation Kit from Stratagene and the PCR Optimisation Kit from Boehringer-Mannheim. Optimising the PCR reaction for direct sequencing is often of greatest benefit when analysis of the same target is required for large sample numbers and the high throughput advantages easily outweigh the time invested in optimisation.
- *PCR using tailed primers* — The impact of general primer design

considerations on the specificity and sensitivity of the PCR reaction have already been discussed in Chapter 5, Section 5.3.2. When designing PCR primers for the generation of a sequencing template, performing amplification using a chimeric primer that consists of the M13 forward or reverse primer sequence at the 5′ end plus the gene-specific portion incorporates the M13 sequence into the PCR product. Incorporating the M13 sequence makes it easy to cycle sequence the PCR products using either the forward (-21M13) or the reverse (M13 Rev.) dye primers, thus avoiding the need to make a custom dye-labelled sequencing primer. This method makes it possible to sequence many different PCR templates using the −21M13 and M13 Rev. sequencing dye primers. However, when using tailed primers the risk of primer dimer formation is greater and additional optimisation may be necessary.

- *Influence of PCR primers and primer dimer on sequencing* — An excess of primer remaining after PCR can act as a primer in the sequencing reaction. During dye primer sequencing the resulting products are not labelled and therefore do not show up during detection. However, they may alter the primer-to-template ratio of the reaction and compromise the dye primer signal obtained. If primer is carried forward from PCR to a dye terminator cycle sequencing reaction, both the forward and reverse PCR primers can act as extension primers. This results in the generation of additional sets of dye-labelled sequencing fragments, making interpretation of the sequencing data virtually impossible. For this reason, direct sequencing of PCR products using dye terminators without any intermediate purification is not generally recommended. If the PCR primers dimerise or concatemerise to create artefact products and the same primer is used for sequencing, the primer can anneal and extend these artefacts. When oligomerisation occurs, the sequence data appear contaminated near the primer owing to the detection of a second set of fragments specific to the PCR primer sequence or its oligomerisation (Figure 10.1). In some cases, these oligomers are long enough or of sufficient molecular weight to persist through PCR product clean-up procedures. Reducing the occurrence of such artefacts during PCR can be achieved using hot start approaches (see Figures 10.2a and 10.2b) or by redesigning the primers to prevent oligomer formation.

- *Product purification method* — In circumstances where multiple targets are to be analysed from multiple samples, the size of the project may preclude the time required to optimise a diverse series of reactions for a direct sequencing approach and the inclusion of a simple post-PCR purification step may be a more desirable option. Numerous methods have been developed for successfully purifying PCR products away from excess reaction components prior to cycle sequencing. Filtration preparations such as Centricon® R-100 microconcentrator columns use an ultrafiltration membrane which has a specific molecular weight threshold and separates single-stranded DNA less than 300 bases and double-stranded DNA less than 125 base pairs away from primers and dNTPs.[7] Carryover

Figure 10.1 *Data produced by dye terminator sequencing of primer-dimer contaminated PCR product*

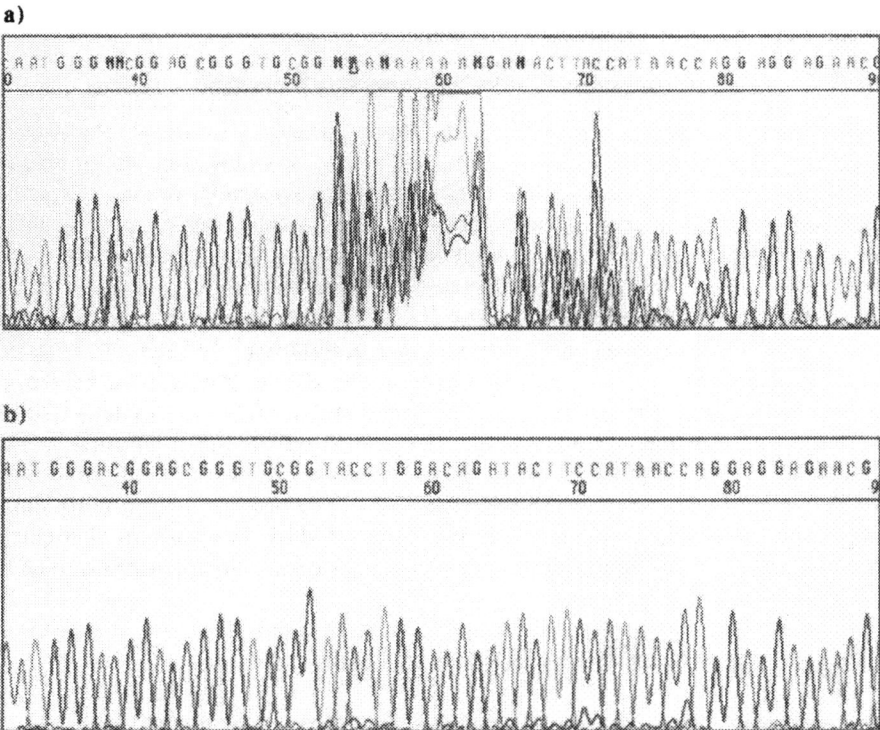

a)

b)

Figure 10.2 *Dye primer sequencing of PCR products generated from the HLA DRB gene using* (a) *Amplitaq® DNA Polymerase and* (b) *Amplitaq Gold® DNA Polymerase, illustrating the improved sequence quality obtainable using an invisible hotstart approach as afforded by Amplitaq Gold®*

reagents can also be removed through a combination of exonuclease I, which degrades residual PCR primers, with shrimp alkaline phosphatase, which dephosphorylates remaining dNTPs. Other commercially available products include QIAquick^{TM} PCR purification columns, which use silica gel to adsorb the DNA in special binding buffers. Alternatively, if the desired PCR contains other amplification products or artefacts, gel purification using low melting temperature agarose (SeaPlaqueR or SeaPrepR) followed by electroelution, glass resin or other standard recovery protocols will provide reliable templates.[8,9] More detailed discussions of purification protocols can be found in a variety of troubleshooting and application guides.

- *Template quantitation* — The importance of a well validated quantitation method has been discussed elsewhere in this manual (Chapter 4) and the same principles hold true for fluorescent sequencing protocols. Of particular note is the importance of rechecking the DNA quantitation following column purification, as it is generally not safe to assume 100% recovery. Accurate quantitation is essential to ensure the addition of optimum amounts of template as overloading a sequencing reaction can be as detrimental as failing to add sufficient quantity for analysis.

10.2.2 The Fluorescent Cycle Sequencing Reaction

When making a decision on which sequencing chemistry to adopt, the overall aim of the sequencing project must be the major consideration. Examples of such projects include *de novo* sequencing in which sequences are compared to databases. In this situation, the priority is often throughput, processing large numbers of samples with quite a high tolerance of ambiguity in the sequence. Another application is the verification of known DNA (checking clone constructs or site-directed mutants) or diagnostic sequencing (comparison of a known wild-type sample against mutants in a population). In the latter case the priority is accuracy of basecall, for example the characterisation of heterozygotes. In addition, factors such as time, cost and whether the system will be suitable for high throughput can also have bearing on the final choice. The availability of pre-formulated reagent kits, optimised to produce consistent and reproducible fluorescent sequencing, aids the user to produce high quality data. However, the quality of the data can still be affected by the choice of chemistry, the amount of template required for sequencing or the instrumentation used to cycle the reaction.

Factors affecting the validity of fluorescent sequencing protocols as a result of the sequencing reaction can be identified as follows.

- *Selection of the appropriate sequencing chemistry* — Sequencing kits can be categorised into two groups depending on where the fluorescent dye label is attached to the sequencing product. The two categories are dye primer and dye terminator, the characteristics of which are summarised in Table 10.1.

Table 10.1 *Advantages and disadvantages of dye primer and dye terminator sequencing chemistries*

Chemistry	Advantages	Disadvantages
Dye primer	Even peak heights Long read lengths	Four-tube reaction Labelled, characterised primers False stops detected Data compressions
Dye terminator	One-tube reaction Unlabelled primers Data compressions very rare False stops not detected	Some variation in peak heights Shorter read lengths

Dye terminator chemistry represents the most popular choice of sequencing format owing to its easy one-tube set-up and the ability to use any unlabelled primer as the sequencing primer. Dye terminators have further advantages in that only those products terminated correctly with a dye-labelled base will be detected by the system. This means that the data do not contain artefacts known as 'stop peaks' (observed with dye primer chemistry) caused by termination of chain growth owing to template secondary structure. Although the peak height pattern is less even than that obtained with dye primers[10] (see Figures 10.3a and 10.3b), this variation provides a useful diagnostic tool for identifying base changes during heterozygote analysis.

The dye primer chemistry approach is commonly used for more specialised applications or for when longer read lengths are required. The more even peak heights facilitate the quantitative detection of multiple base locations as necessitated during some types of heterozygote detection and maximises the chance of detection even when sample amounts are limiting. This chemistry also has the advantage that it does not require the complete removal of PCR primers for direct sequencing. For the more specialised applications, the requirement for accurate base detection generally overrides the potential drawbacks of dye primer reactions as, ultimately, the quality and accuracy of the results are of primary importance.

Compression is an artefact of sequencing data which may lead to the squashing up of the data peaks or, at worst, adjacent peaks swapping position.[11] Compression is caused by the reaction product attaining some secondary structure as it moves through the gel and is prevented by the use of an analogue of dGTP in dye primer chemistry. In the case of dye terminator sequencing, the analogue used is inosine which is extremely efficient at preventing compressions. Compression is now largely overcome by running the gel at elevated temperatures (Section 10.2.5).

By weighing the requirement of the application against the advantages and disadvantages of the available chemistries, the experienced user should

a)

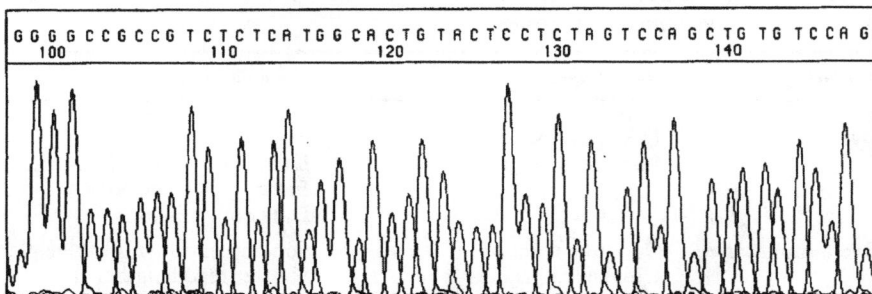

b)

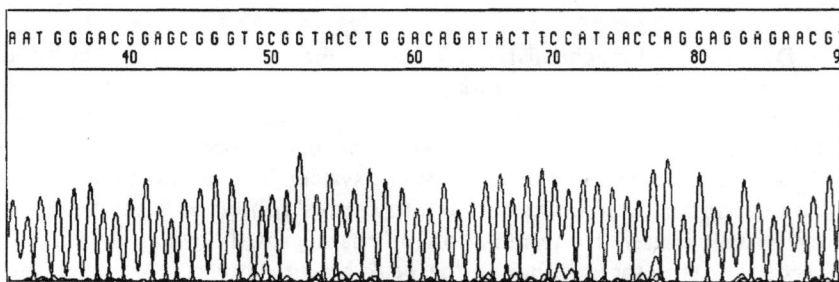

Figure 10.3 *Examples of the type of result produced by* (a) *dye terminator chemistry and* (b) *dye primer chemistry*

be able to decide upon an approach that will maximise the chances of a successful result on the majority of templates.

• *Optimised reagent kit format* — Some kits contain a preformulated mix which requires only the addition of DNA and, in the case of terminator chemistries, the primer. Such kits contain the correct balance of reagents necessary to perform a successful reaction on the majority of templates commonly encountered. Should the user wish to adapt certain aspects of the protocol, kits are available with the component reagents separated to allow user discretion. Any alterations to the recommended protocol should be vigorously investigated to ensure no detrimental effects on the quality of the final result. Choosing the type of kit designed for use on the detection platform employed will further enhance the reliability of the system.

• *Input template concentration* — Too much starting template will result in too much fluorescence being produced. This in turn will cause problems in the ability of the instrument to resolve the spectral overlap of the dyes (Figure 10.4a). It will also impact upon the ability of the system to resolve finely the peaks of the larger sequencing products; therefore the overall length of acceptable data will be reduced. Insufficient template will have the reverse effect of producing insufficient fluorescence, thus reducing the

a)

b)

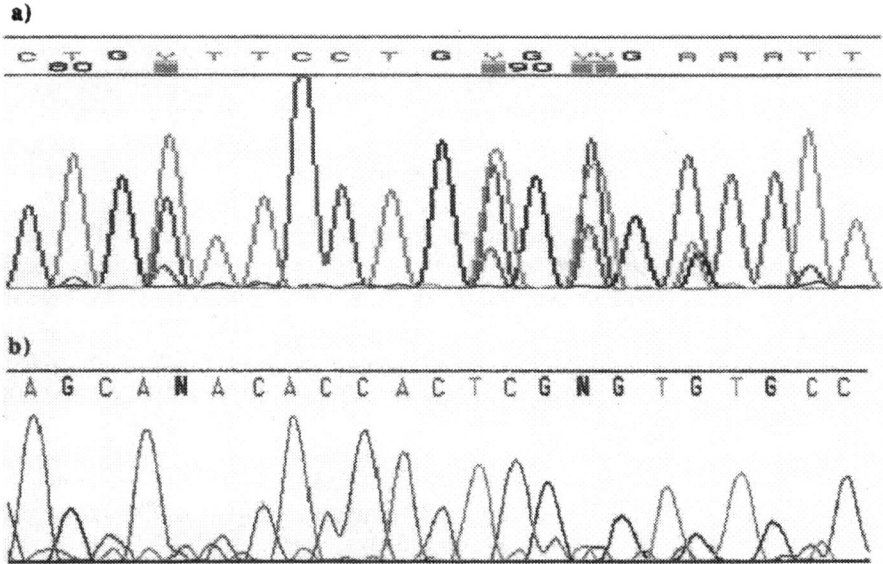

Figure 10.4 *Dye terminator sequencing showing* (a) *'pull-up' peaks caused by an excess of starting template and* (b) *insufficient starting template giving low signal strength and noisy data resulting in ambiguity calls*

level of signal (Figure 10.4b). This will only begin to impact upon the data quality when the system is unable to differentiate between what is signal and what is background. At worst, with no signal to detect, the software will not be able to analyse the data. However, even at levels where the signal can be detected, the data will still appear poor due to the low signal-to-noise ratio. Minimum and maximum template requirements should be determined for whichever template is in use, bearing in the mind the target copy number expected to be present. PCR products provide a target-rich template and, as such, template amounts are generally quite small (refer to relevant kit protocols for more detailed discussions of input DNA concentrations). Normal plasmid and genomic DNA preparations require slightly increased template amounts, which should be increased still further if the template is large or the target is known to be a rare copy. For a valid protocol the correct amount of starting template should maximise the chances of obtaining a successful result and reflect both the type of template and the density of target within the sample.

• *Input primer concentration* — Adding too much or too little primer to the sequencing reaction can result in insufficient signal for analysis, or a sequence with reduced read length, respectively. Reagent manufacturers generally recommend a primer concentration which will work well for the majority of templates. Slight adjustments of this concentration may improve data quality but optimisation of primer concentration is generally less critical than template amount as the efficiency of the primer is determined more by its design than by slight variations in the amount present.

- *Primer design* — Considerations relating to PCR primers are also valid when designing sequencing primers for dye terminator reactions or custom dye primers. Poor primer design generally results in one of two outcomes:

  - Poor priming resulting in a weak signal can be a consequence of too low a melting temperature, secondary structure within the primer, particularly at the 3' end, or secondary structure of the template at the primer binding site.
  - Good signal strength but noisy data can result from a lack of specificity in the primer, resulting in extra peaks, or an impure primer producing sequence shadows.

  There is a further consideration when designing custom dye primers. Owing to the nature of some of the dyes attached to the primers, the fragment mobility on a gel is affected over the first 50–75 bases of the sequencing ladder by interaction of the dye label with the first five bases of the sequencing primer. The sequencing analysis software installs mobility shift files for each of the standard dye primers and uses the files to correct the mobility of the fragments labelled with those primers.

  To adjust for mobility shifts if a custom primer is made, the first five bases of the M13 reverse primer (5'-CAGGA-3') must be added during the synthesis to the 5' end of the primer. The reverse primer mobility file can then be used to analyse the sample.

- *Thermal cycler considerations* — For cycle sequencing reactions the choice of a robust and reliable instrument can significantly enhance the quality of the data generated. Should a thermal cycling block fail to produce equal thermal cycling profiles both across the block and between runs, individual samples or sample batches may be subjected to fluctuating conditions, resulting in inconsistencies in the data (discussed in Chapter 5).

10.2.3   Template Sequence Composition

It is unreasonable to expect any given chemistry to perform optimally on every type of template and consideration should be given to the appropriate protocol modifications required to deal with a variety of template-specific idiosyncrasies. Sequence compositions which may affect the validity of fluorescent DNA sequencing protocols can be identified under the following headings.

- *GC-rich templates* — GC base composition varies from 25 to 75% depending on the organism, but any sequence may contain a region with high GC content. In such sequences the single-stranded template is able to form very stable interactions with itself. This secondary structure, if severe enough, will prevent the polymerase from processing along the template, causing it to drop off and the chain elongation to stop (Figure 10.5). The symptoms of this are different, depending upon the chemistry used. Dye primer chemistry will show a large stop peak with little or no data following it. Dye terminator data will simply fall to very low levels or stop completely. The data quality can be improved by modifications aimed at forcing enough of the polymerase enzymes to process through the region, thereby generating sufficient data to be analysed. The most popular approach is to add DMSO to a final concentration of 5%. This has the effect of reducing the effective melting temperature of the DNA, allowing the enzyme to force itself through the secondary structure. The limitation of this approach is that, as well as melting the DNA, the DMSO also partially deactivates the polymerase; it may therefore be necessary to supplement the amount of enzyme in the reaction.

- *Templates containing secondary structure* — Inverted repeats and palindromes can cause problems during sequencing due to their tendency to form secondary structures which, as in the case of GC-rich templates, impairs the ability of the enzyme to read through the region, as demonstated in Figure 10.5. Using a 5% DMSO or similar protocol modification or raising the denaturation temperature may solve the problem.

- *AT-rich templates* — The potential problems presented by AT-rich sequences arise more commonly during data analysis as it is generally easier to sequence through AT-rich regions than their GC counterparts. Uneven peak heights resulting from terminator protocols may be more exaggerated when large numbers of A and T bases are present. A thorough understanding of the peak patterns associated with particular enzymes and chemistries should allow the experienced user to interpret such events when analysing the data. However, the more even peak heights generated by dye primer chemistry may provide an alternative approach.

- *Templates containing repetitive DNA* — Short repeats can generally be sequenced without difficulty, but extended regions of repeats, particularly strings of a single base, may result in enzyme slippage, leading to the production of a shadow sequence. The *Taq* sequencing enzyme is not wholly processive, which means that it takes several enzyme molecules to complete a full-length extension product. Each molecule adds 50–60 nucleotides before dissociating and allowing another molecule to continue. As the enzyme dissociates, the product and template strands are thought to 'breathe'. As the next molecule joins on, the strands are brought together, but the presence of short repeat motifs may cause the strands to reanneal out of step. This results in product copies several repeat units shorter than the original strand, visible as shadows under the original sequence which, at worst, may obscure the correct sequence, making unambiguous interpretation difficult (see Figure 10.6). Problems with G and C homopolymer repeats may occur with less than 10 bases, whereas for A and T, strings of up to 25 or 30 bases can be sequenced effectively before the enzyme gets into difficulty.

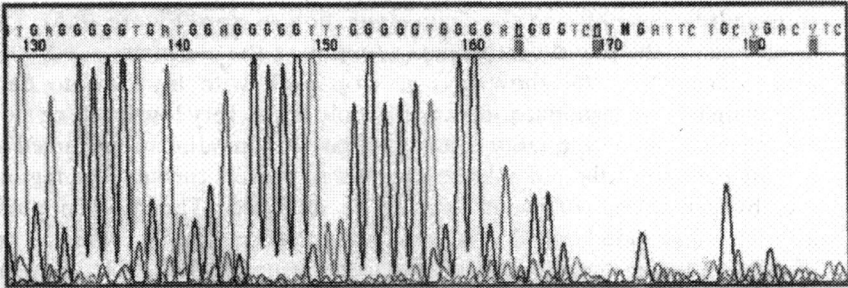

Figure 10.5 *Secondary structure impairs the ability of the enzyme to generate full length extension products, characterised by a sudden reduction in signal*

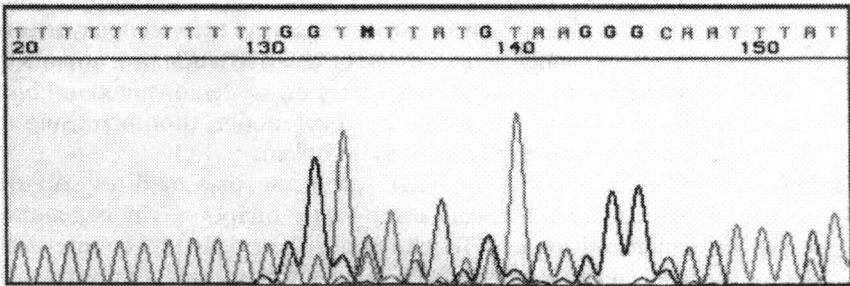

Figure 10.6 *Characteristic stutter pattern resulting from enzyme slippage in homopolymer repeat regions*

10.2.4 Post-reaction Purification

Failure to purify sequencing products adequately, following the sequencing reaction, can result in sub-optimal data quality due to the interference of reaction components as detected by downstream data handling packages. In particular, dye molecules carried forward from the sequencing reaction to the detection platform, in the form of excess labelled dNTPs or primers, may obscure the beginning of the sequence such that a substantial portion of the data is lost.

Some factors affecting the validity of fluorescent DNA sequencing protocols due to standard purification methods have already been discussed in relation to precipitation protocols following extraction (see Chapter 3, Section 3.4.6). Other factors, unique to the precipitation of sequencing reactions, can be identified as follows.

- *Type of purification* — Simple ethanol precipitation protocols recommended by reagent manufacturers offer an efficient and cost effective means of cleaning up the reactions, provided the method is performed well. Unlike standard ethanol precipitation methods, it is worth noting that overloading the final reaction volume with ethanol or incubating the precipitation at $-20\,^{\circ}\text{C}$ to encourage DNA precipitation can have detrimental effects on the final data as both promote the precipitation of dye molecules along with the DNA. Ethanol precipitation can be laborious when handling large numbers of samples so, as an alternative, several commercial suppliers offer reaction clean-up columns which operate on a size exclusion basis. These allow sequencing products to be retained whilst eliminating excess contaminants. Though more costly, such preparations provide a more rapid means of sample purification for high throughput applications.

10.2.5   Gel Preparation

Ultimately, a successful and well designed sequencing protocol is only as effective as the detection method employed. Some instrument platforms employ proprietary separation matrices preformulated by the instrument manufacturer. Provided the recommended protocol is followed, the need to optimise the separation matrix formulation and preparation may not arise. For platforms employing slab gel electrophoresis as the means of separation there is a requirement for a uniformity in gel characteristics, ensuring that the fragments display the same mobility from run-to-run. There is a broad window of acceptability, determined by the ability of the software to interpret the speed of the fragments through the gel. If, however, the gel make-up means that the fragments pass through the gel too quickly or too slowly, so falling outside this window, the software will not be able to analyse the data effectively. It is also important that the background fluorescence of the gel is kept to a minimum; this is partially ensured by the use of reagent grade chemicals.

Factors affecting the validity of fluorescent DNA sequencing protocols as a result of gel preparation techniques can be identified as listed below

- *Glass plate cleanliness* — The cleanliness of the glass plates is of paramount importance when using fluorescent detection, as any fluorescent contaminant on the plate surface may interfere with or obscure the sample fluorescence, thereby confusing data interpretation. Washing protocols based around minimal use of specialised detergents such as Alconox®, together with avoidance of normal tap water in favour of a high quality deionised water supply, generally produce the best results. A periodic stringent wash protocol involving nitric acid and sodium hydroxide solution may help to prevent contaminant accumulation on the plate surface.
- *Reagent quality* — The quality of the gel relies to a large degree upon the quality of reagent. The use of ultrapure grade constituents will help to ensure the least possible risk of fluorescent contamination arising from the reagent source.
- *Gel recipe and preparation method* — Specific gel recipes are available from a variety of sources, including acrylamide and instrument manufacturers. High quality gels are produced through the dilution of a concentrated bis/acrylamide solution to the required percentage, followed by reagent

combination just prior to pouring. The inclusion of deionising and degassing steps at appropriate junctures optimises the polymerisation process and consequently maximises the detection of successful sequencing reactions. Some manufacturers supply bis/acrylamide solutions premixed with all the necessary components of a standard denaturing sequencing gel, with the exception of one or more polymerisation catalysts. Such products generally result in a lower quality gel than fresh reagent combination methods, but for applications requiring high throughput, the time investment issues arising from making fresh gels may impose too significant a penalty and such premixes offer an alternative.

- *Gel percentage* — Gel percentage depends on the instrument platform employed, the bis/acrylamide ratio and the application for which data are required. Instrument manufacturers recommend a general bis/acrylamide ratio and gel percentage guidelines suitable for a broad spectrum of applications. The most common gel percentages range from 4 to 6% based on a 19:1 acrylamide:bis ratio. Users may find that trying slight modifications to the recommended gel percentages or acrylamide:bis ratios may improve the resolution and read length for some templates. It should be remembered, however, that the software requires the DNA to move through the gel at a certain speed. Changing either the acrylamide:bis(acrylamide) ratio or the percentage of the gel will have an effect on the mobility of the fragments. Users making any alterations to the gel recipe must ensure the mobility of the DNA is not significantly altered.

- *Polymerisation rate and efficiency* — Uneven gel polymerisation can result in lane migration and tracking failure. Pouring a gel in a warm laboratory environment or with warm acrylamide can cause polymerisation to occur too quickly, changing the characteristics of the matrix and causing the gel to run too slowly.

- *Gel age* — For maximum performance, gels should, on average, be allowed to polymerise for 2 h and then run as soon as possible. Running a gel too soon after pouring or leaving a gel too long can lead to poor resolution and short read lengths. Depending on the instrument platform and gel recipe used, gels as old as 8 h may begin to show signs of degradation. Gels older than 24 h should be discarded.

- *Gel running* — Incorrect buffer measurement or its accidental deionisation can cause the gel to run anomalously, producing a fragment migration rate outside the limits accepted by the analysis software. Compressions can now largely be overcome by the use of gel-heating devices, which maintain a constant temperature of $\sim 50\,^{\circ}\text{C}$.[12]

As a general rule, sub-optimal gels will result in sub-optimal data. For a valid protocol, the data quality required should be considered with reference to the various stages of gel preparation; if necessary, areas where compromise may be tolerated should be identified. This should enable the experienced user to develop a protocol which satisfies the major requirements of the application.

10.2.6 Data Analysis and Interpretation

The ease and accuracy of interpreting fluorescent sequencing data has been substantially improved over recent years through the combined introduction of automated fluorescent detection and computer mediated data analysis. Removing the subjectivity of human judgement for the majority of the process plays a substantial part in reducing the possibility of error and improving the validity of the technique, even for high stringency applications. Developing a valid protocol where reliance is placed on such automated systems requires considerations related more to the degree of competency of the user in operating the system rather than in the actual design of the package itself.

Factors affecting the validity of fluorescent DNA sequencing protocols as a result of data analysis and interpretation can be identified as follows.

- *Effective training and understanding of the packages involved* — Correct manipulation of the software and an understanding of the impact of different parameters on the appearance of the final result will ensure that a valid analysis protocol is developed. Failure to understand the impact of inappropriate analysis parameters on data quality may result in sub-optimal or inaccurate data being produced, hence invalidating the protocol.
- *Software assessment* — For some high stringency applications it may be necessary to assess the software based on the results of a set of existing standard samples analysed under existing protocols. This should ensure that, although analysis techniques may develop in terms of their sophistication and discrimination, consistency is maintained.

10.3 Summary

The different stages constituting an automated fluorescent DNA sequencing protocol, as discussed in this chapter, are designed as a guide to adopting a sequencing approach of this type. To ensure the final protocol is valid, each stage of the process should be evaluated to ensure the results produced are informative with regard to the applications. Considerations such as time, cost and stringency are important in selecting the correct approach and determining which areas of the process require particular attention and where compromises may be tolerated. A variety of publications are available to aid in this process, a selection of which are detailed in the associated references section.[13-17] Theoretical research coupled with empirical determination of the most appropriate combination of steps will result in high quality results suitable for the purpose for which they are intended.

Applied Biosystems, PE, PE Applied Biosystems are trademarks and ABI Prism and Perkin Elmer are registered trademarks of the Perkin-Elmer Corporation.

Amplitaq and Amplitaq Gold are registered trademarks of Molecular Systems.

10.4   References

1. Lawyer, F. C., Stoffel, S., Saiki, R. K., Myambo, K., Drummond, R. and Gelfand, D. H. 1989. Isolation, characterisation and expression in *E. coli* of the DNA polymerase gene from the extreme thermophile *T. aquaticus. J. Biol. Chem.* **264**: 6427–6437.
2. Lawyer, F. C., Stoffel, S., Saiki, R. K., Chang, S. Y., Landre, P. A. and Abramson, R. D. 1993. High-level expression, purification and enzymatic characterisation of full length *T. aquaticus* DNA polymerase and a truncated form deficient in 5′ to 3′ exonuclease activity. *PCR Methods Applications* **2**: 275–287.
3. QIAGEN GmbH. 1995. QIAGEN 1995 Product Guide (catalog). QIAGEN, Hilden, Germany.
4. The Perkin Elmer Corporation. 1995–1996. Perkin Elmer PCR Systems, Reagents and Consumables (catalog). The Perkin Elmer Corporation Applied Biosystems Division, Foster City, CA.
5. Innis, M. A. and Gelfand, D. H. 1990. Optimisation of PCRs. In: PCR Protocols: A Guide to Methods and Applications (eds. Innis, M. A., Gelfand, D. H., Sninsky, J. J. and White, T. J.), chap. 1. Academic Press, San Diego.
6. Newton, C. R. and Graham, A. 1994. PCR (eds. Graham, J. M. and Billington, D.). Bios Scientific, Oxford.
7. Hanke, M. and Wink, M. 1994. Direct DNA sequencing of PCR-amplified factor inserts following enzymatic degradation of primer and dNTPs. *BioTechniques* **17**: 858–860.
8. Zhen, L. and Swank, R. T. 1993. A simple and high yield method for recovering DNA from agarose gels. *BioTechniques* **14**: 894–898.
9. Werle, E., Schneider, C., Rennet, M., Völker, M. and Fiehn, W. 1994. Convenient, single-step, one-tube purification of PCR products for direct sequencing. *Nucleic Acids Res.* **22**: 4354–4355.
10. Parker, L. T., Deng, Q., Zakeri, H., Carlson, C., Nickerson, D. A. and Kwok, P. Y. 1995. Peak height variations in automated sequencing of PCR products using *Taq* dye-terminator chemistry. *BioTechniques* **19**: 116.
11. Kapelner, S. N., Turner, R. T., Sarkar, G. and Bolander, M. E. 1994. Deletion mutation can be an unsuspected gel artefact. *BioTechniques* **17**: 64–67.
12. Mardis, E. R., Panussis, D. A., Weinstock, L. A. and Wilson, R. K. 1995. Resistance heating device reduces gel mobility compressions in automated fluorescent sequencing. *BioTechniques* **18**: 622–624.
13. Gibbs, R. A., Nguyen, P. N., Edwards, A., Civitello, A. B. and Caskey, C. T. 1990. Multiplex DNA deletion detection and exon sequencing of the hypoxanthine phosphoribosyltransferase gene in Lesch–Nyhan families. *Genomics* **7**: 235–244.
14. Larder, B. A., Kohli, A., Kellam, P., Kemp, S. D., Kronick, M. and Henfrey, R. D. 1993. Quantitative detection of HIV-1 drug resistance mutations by automated DNA sequencing. *Nature* **365**: 671–673.
15. Leren, T. P., Rødningen, O. K., Røsby, O., Solberg, K. and Berg, K. 1993. Screening for point mutations by semi-automated DNA sequencing using SequenaseR and magnetic beads. *BioTechniques* **14**: 618–623.
16. Petersdorf, E. W. and Hansen, J. A. 1995. A comprehensive approach for typing the alleles of the HLA-B locus by automated sequencing. *Tissue Antigens* **46**: 73.
17. Bronner, C. E., Baker, S. M., Morrison, P. T., Warren, G., Smith, L. G., Lescoe, M. K., Kane, M., Earabino, C., Lipford, J., Lindblom, A., Tannergård, P., Bollag, R. J., Godwin, A. R., Ward, D. C., Nordenskjøld, M., Fishel, R., Kolodner, R. and Liskay, R. M. 1994. Mutation in the DNA mismatch repair gene homologue hMLH1 is associated with heredity non-polyposis colon cancer. *Nature* **368**: 258–261.

Appendix: Glossary of Terms

3′ end (3′ terminus). Amino terminus of a nucleic acid strand.

5′ end (5′ terminus). Hydroxyl terminus of a nucleic acid strand.

Adenine (A). A purine base found in DNA or RNA.

Adulteration. The act of lowering the quality of a material by the addition of an inferior or undesirable substitute.

Allele. One of several variants of a gene at a given locus on a chromosome.

Allelic ladder. A DNA size marker used, for example, in STR profiling, consisting of the various polymorphic forms (sizes) of each allele.

Amplicon. A PCR amplified DNA fragment.

Amplification. The process of increasing the number of copies of a specific predetermined nucleic acid sequence in, for example, a PCR.

AmpliTaq GOLD. A variant of *Taq* DNA polymerase which requires thermal activation. Useful in the prevention of mis-priming events which may occur with *Taq* at sub-optimal temperatures.

Analytical molecular biology. The science of analysing and understanding biological materials and processes at the molecular level.

Anion. A negatively charged ion or molecule.

Annealing. The process by which two complementary single strands of a nucleic acid interact.

Antibody. A specific glycoprotein produced by the immune system of vertebrates in response to exposure to a foreign substance (antigen). Antibodies are capable of binding specifically and with great affinity to their target antigen.

Antigen. A substance that elicits the formation of antibodies in a vertebrate host.

Artefact. In the context of PCR, an artefact is a non-specific amplification product such as a mis-primed amplicon or primer dimer.

Autoradiography. The technique by which X-ray film is exposed to a radioactive source to generate an image that localises the radioactive components present in a biological specimen, gel or membrane.

Bacteriophage. A virus capable of infecting bacteria.

Band. Term used to describe a discrete DNA fragment following gel electrophoresis and visualisation by UV illumination or autoradiography.

Base. Purine or pyrimidine component of a nucleotide. For DNA, these are A and G (purines), C and T (pyrimidines). For RNA, T is replaced by U.

Base pair (bp). Association of complementary nucleotides (A with T or U and C with G) on opposing antiparallel strands of a nucleic acid double helix by means of hydrogen bonds. The term base pair is also used to indicate the size of a nucleic acid fragment, e.g. a 100 bp DNA fragment is a double helical DNA comprising 100 nucleotides per strand.

Basecall. Process by which ABI fluorescent sequencing software assigns a specific base to a spectral peak.

Biosciences. Biological sciences, including biotechnology, molecular biology, biochemistry, cell biology, etc.

Biotechnology. The scientific manipulation of living organisms or their components at the molecular level to enhance desirable traits or produce useful products.

Blank control. A solution or mixture of reagents that excludes the analyte under analysis, routinely used to eliminate any signals arising from background interference.

Blind control. A sample of origin unknown to the analyst, used to test both the specificity of a particular assay and competency of the analyst.

Cation. A positively charged ion or molecule.

Cell membrane. The structure surrounding a cell that consists of a lipid bilayer and proteins.

Chaotropic agent. A protein denaturant, used for the solubilisation of membranes.

Chromosome. The structure of DNA and associated proteins in the nucleus of a eukaryotic cell, and the circular DNA of a prokaryotic cell that carries genetic information.

Codon. A sequence of three adjacent nucleotides (triplet) in mRNA that specifies and is translated into a particular amino acid or termination signal.

Cofactor. A small non-protein compound or ion that is required for enzyme activity.

Colony. A population of cells, arising from a single cell, growing on a solid medium.

Competitive PCR. A variant of standard PCR involving co-amplification of two target DNA sequences using the same primer pair. Used for quantitative purposes.

Complementary. The specific pairing, via hydrogen bonds, of A to T (or U in RNA) and G to C in opposite antiparallel strands of DNA or RNA.

Complementary DNA (cDNA). A single-stranded DNA synthesised from mRNA by reverse transcription.

Complementary strand. The nucleic acid strand that has a complementary base sequence to another strand, e.g. $^{5'}$ATCG$^{3'}$ is the complementary strand of $^{3'}$TAGC$^{5'}$.

Consensus sequence. The 'average' sequence calculated from a group of highly similar sequences.

Contamination. In PCR, this refers to the unintentional introduction of nucleic acids into the reaction mixture, via aerosols or poor working practice. Contaminating nucleic acids can be non-target sequences from an extraneous source, post-PCR amplicons or target DNA cross-contamination between PCR tubes.

Cultivar. A variety of plant produced through selective breeding by humans and maintained by cultivation.

Culture. Cultivation of microbial cells in or on nutrient medium.

Cycle sequencing. PCR-based method of nucleic acid sequencing.

Cytosine (C). A pyrimidine base found in DNA or RNA.

Denaturation of nucleic acids. The dissociation, chemically or thermally, of the complementary strands of a double-stranded nucleic acid into single strands.

Denaturation of proteins. The disruption of the native conformation of a protein to some other form, leading to a loss of biological activity.

Deoxyribonucleotide. The monomer of DNA that consists of deoxyribose-phosphate esters with each deoxyribose unit carrying a base (A, T, C or G).

Deoxyribose. Five-carbon sugar component of DNA.

Detergent. Naturally occurring or synthetic amphipathic substance that disrupts cell and organelle membranes by intercalation into the membrane matrix and solubilisation of the component lipids and proteins. Can be **ionic** (with charge, e.g. SDS) or **non-ionic** (without charge, e.g. Triton X-100, Tween 20)

Dideoxy sequencing. Chain termination method for DNA sequencing devised by Sanger. Involves the use of dideoxynucleoside triphosphate chain terminators (deoxyribonucleotide analogues that do not have a hydroxyl group at the 3′ position).

DNA (deoxyribonucleic acid). A deoxyribonucleotide polymer, comprising a deoxyribose-phosphate backbone with base side chains (A, T, C, G), which provides the genetic information of most living organisms.

DNA cloning. The *in vivo* process by which genetic sequences are replicated outside the original host organism.

DNA fingerprinting. A technique devised by Jefferys and used to generate individual-specific patterns of polymorphic fragments at multiple loci, resulting from restriction endonuclease digestion of genomic DNA.

DNA polymerase. An enzyme that catalyses the synthesis of a new complementary DNA strand from a single stranded DNA template.

DNA Polymorphism. A variation in the sequence of DNA between individuals.

DNA profiling. A collective term for the variety of techniques used to generate specific patterns of polymorphic DNA fragments for identification purposes.

DNase. An exonuclease that catalyses the internal hydrolysis of DNA.

Dot/slot blot. Hybridisation-based technique used for the detection and/or quantification of a specific target nucleic acid, in which the denatured sample is applied directly to a membrane as a dot or slot.

Double helix. A spiral arrangement of two intertwining anti-parallel complementary DNA strands held together by hydrogen bonds between side chain bases.

Duplex. See double helix.

E. coli (*Escherichia coli***).** A Gram-negative bacterium of the *Enterobacteriaceae* commonly found in the vertebrate intestine. Frequently used in molecular biology as a host for DNA cloning.

Electrophoresis. A commonly used technique for separating charged molecules (e.g. proteins, nucleic acids) based on their differential mobility in an electric field.

Endonuclease. An enzyme that cleaves phosphodiester bonds within a nucleic acid strand. Cleavage of the phosphodiester bond can be on the 3′-phosphate side (3′–5′ endonuclease) or on the 5′-phosphate side (5′–3′ endonuclease).

Enzyme. A bioactive protein that catalyses biochemical reactions within a living cell. A variety of enzymes are exploited in the biosciences for the *in vitro* catalysis of a wide range of biological reactions.

Ethidium bromide (EtBr). A fluorescent dye that intercalates between the stacked bases of nucleic acids, causing the nucleic acid to fluoresce under UV light. EtBr is commonly used for the visualisation of discrete DNA fragments on an agarose gel following electrophoretic separation.

Eukaryote. Organism composed of cells which possess a membrane-bound nucleus and cell organelles, e.g. plants, fungi, animals.

Exonuclease. An enzyme that catalyses the sequential removal of nucleotides from either the 3′- or 5′-ends of a polynucleotide chain. Exonucleolytic cleavage can proceed in the 5′ to 3′ direction (5′–3′ exonuclease) or in the 3′ to 5′ direction (3′–5′ exonuclease).

False negative. A result that is incorrectly recorded as negative because of errors in the procedure. In PCR, this often refers to amplification failure as a result of the presence of PCR inhibitors in the target sample or reaction.

False positive. A result that is incorrectly recorded as positive because of errors in the procedure. In PCR, this is often a result of contamination or low primer specificity.

Familial analysis. Investigation of genetic relationships (e.g. parentage) within a family or group of individuals.

Fidelity. The accuracy of a DNA polymerase in synthesising the complementary strand to the nucleic acid template during primer extension.

Fluorescence. The emission of light at a specific wavelength by an excited molecule in the process of making the transition from the excited to the ground state.

Fluorophore. A fluorescent molecule, commonly used to label oligonucleotides for the direct visualisation of nucleic acids.

Free magnesium. In a PCR reaction, this refers to the proportion of magnesium not bound to dNTPs in the reaction mix and hence available for the *Taq* DNA polymerase.

GC content. The percentage of G and C residues in a nucleic acid or genome.

Gene. A sequence of nucleotides that codes for a functional protein product.

Genetic code. The relationship between the sequence of nucleotide triplets (codons) in a gene and the amino acids in the protein they encode.

Genetic disease. An inherited disease caused by changes (mutations) in the genetic material of an individual.

Genome. The entire chromosomal content of an organism.

Genotype. The genetic characteristics of an organism.

Glycoprotein. A protein linked to an oligosaccharide or polysaccharide.

Guanine (G). A purine base found in DNA or RNA.

Hairpin. A double helical region on a single-stranded DNA or RNA molecule formed by base pairing between adjacent complementary sequences.

Heme/haem. An iron–porphyrin complex that forms the oxygen-binding portion of haemoglobin.

Heterozygote. An organism that carries two different alleles for a given sequence or gene.

Homology. The degree of similarity between two distinct nucleotide sequences.

Homopolymer. A polymer that consists of repeating units of the same monomer.

Homozygote. An organism that carries two identical alleles for a given sequence or gene.

Hot start PCR. A variant of standard PCR which involves the exclusion of an essential component of the reaction mixture (e.g. *Taq* polymerase or magnesium) until all target nucleic acids have been fully denatured. Hot start PCR serves to enhance the specificity of amplification by minimising mis-priming events that can occur at sub-optimal annealing temperatures.

Housekeeping genes. Genes that provide the basic functions required for cell survival, such as those encoding enzymes for biochemical processes.

Hybridisation. The process of complementary base pairing between two single strands of nucleic acid. Hybridisation can occur between DNA–DNA, DNA–RNA and RNA–RNA, in solution (solution hybridisation) or on a membrane (membrane hybridisation).

Hydrogen bond. A weak interaction between a hydrogen atom and an electronegative atom such as oxygen or nitrogen. The nucleic acid double helical structure is held together by hydrogen bonds formed between $T + A$ and $C + G$ in DNA, and $U + A$ and $C + G$ in RNA.

In vitro. Literally meaning 'in glass'. Describes biological reactions occurring in a test tube as opposed to biochemical reactions taking place within the living cell (*in vivo*).

Inverted repeat. Two copies of the same DNA sequence orientated in opposite directions on the same molecule.

Kilobase/kilobase pair (kb/kbp). 1000 base/base pairs of nucleic acid.

Klenow fragment. The polypeptide fragment of *E. coli* DNA polymerase I which is heat labile and possesses both DNA polymerase and 3′ to 5′ exonuclease activity.

Lambda phage. A double-stranded DNA bacteriophage.

Linkage. The tendency of different genes to be inherited together as a result of their proximity on the same chromosome.

Lipid. A hydrophobic organic molecule, e.g. fats, fatty acids, steroids.

Locus (plural, loci). The position on a chromosome where a particular gene or sequence resides.

Long PCR. A variant of standard PCR, allowing amplification of long nucleic acid targets, up to 40 kb.

M13. A filamentous bacteriophage containing single-stranded DNA, commonly used as a vector into which specific DNA sequences can be inserted for DNA sequencing.

Medium (plural, **media**). Solid or liquid nutrient preparation used for the propagation and culture of microorganisms and cell lines.

Melting temperature (T_m). The temperature, during denaturation, at which the transition from double-stranded DNA to single-stranded DNA is 50% complete.

Membrane. Inert solid support (nylon or nitrocellulose) commonly employed for the immobilisation of nucleic acids in Southern, Northern, dot and slot blot techniques.

Messenger RNA (mRNA). The RNA, transcribed from DNA by RNA polymerase, that specifies the amino acid sequence for a polypeptide.

Microsatellite. Refer to **STR**.

Microtitre plate. 96-well plate commonly used for ELISA-based assays.

Mimic. A competitive PCR fragment of similar size and base composition to the PCR target amplicon. A mimic is co-amplified with the target molecule using the same primer pair and is used for the quantification of the DNA target.

Mismatch. An imperfect base pairing between two nucleic acid strands.

Mis-priming. The imperfect annealing of a PCR primer to a non-target region on the nucleic acid template.

Mitochondrial DNA (mt DNA). The maternally inherited genetic material of the mitochondrion.

Molecular size markers/molecular weight marker (MWM). DNA fragments of known size used to calibrate an electrophoretic gel, e.g. 100 bp DNA ladder.

Multiplex PCR (mPCR). A variant of standard PCR where multiple sequences are simultaneously amplified in a single reaction, using a distinct primer set for each target.

Mutation. A sequence change in a nucleic acid which can be genetically inheritable.

Negative control. In PCR, this is a reaction in which target DNA is excluded. The generation of amplicon in the negative control is indicative of contamination.

Nested PCR. A variant of standard PCR involving amplification of a PCR product with a second primer pair, to improve the specificity of amplification.

Northern blotting. The RNA equivalent of **Southern blotting**.

Nuclease. An enzyme (**DNase** or **RNase**) that cleaves phosphodiester bonds of a nucleic acid (DNA or RNA, respectively).

Nucleic acid. Long linear polymer of ribonucleotides (RNA) or deoxyribonucleotides (DNA).

Nucleoside. A **nucleotide** lacking the phosphate group.

Nucleotide. The base–sugar–phosphate monomer of a nucleic acid molecule.

Oligonucleotide. A short (usually synthetic) fragment of single-stranded nucleic acid.

Organelle. A subcellular membrane-bounded body with a well defined function.

Palindrome. A sequence of bases that reads the same in both 5′–3′ and 3′–5′ directions on opposite strands of the DNA duplex, e.g. GAATTC.

PCR enhancer. A substance which results in an increase in product yield or specificity when added to a PCR reaction.

PCR inhibitor. A substance which results in a decrease in product yield or specificity when added to a PCR reaction.

Pfu **DNA polymerase.** A thermostable DNA polymerase isolated from the thermophilic bacteria *Pyrococcus furiosus*, which possesses 3′ to 5′ exonuclease activity (**proofreading** ability).

pGEM. A plasmid commonly used for DNA cloning.

Phage. See **bacteriophage**.

Phosphodiester bond. Linkage between the 5′ phosphate and 3′ hydroxyl groups of adjacent individual nucleotide residues in a nucleic acid.

Phylogenetics. The classification system based on the evolutionary relationships of organisms.

Plaque. A clear spot on a lawn of bacterial cells, produced by phage-induced bacterial cell lysis.

Plasmid. An extrachromosomal circular DNA capable of self-replication in bacteria.

Plateau effect. Cessation of exponential amplification during PCR, owing to the depletion of one or more of the reagents.

Polymerase. An enzyme that catalyses the synthesis of a polymer from monomers. See also **DNA polymerase** and **RNA polymerase**.

Polymerase chain reaction (PCR). An *in vitro* DNA polymerase-based method of DNA amplification in which repeated cycles of DNA denaturation, primer annealing and extension by *Taq* DNA polymerase are used to amplify exponentially the number of copies of a target DNA sequence.

Polypeptide. A polymer of amino acid monomers joined by peptide (amide) bonds.

Positive control. In PCR, a reaction containing target DNA which is known to amplify under the conditions being employed.

Primer. A short synthetic sequence of single-stranded DNA, complementary to regions flanking the target sequence for PCR which serves as the starting point for complementary strand synthesis.

Primer annealing. The hybridisation of a primer to a single-stranded template nucleic acid during PCR.

Primer dimer. PCR artefact resulting from the annealing and subsequent amplification of a primer pair during the course of amplification.

Primer extension. Synthesis of the complementary strand from the 3′ hydroxyl end of a primer that is annealed to a single-stranded template, during PCR.

Probe. A nucleic acid molecule, commonly used to identify (by hybridisation) complementary sequences in a single-stranded target nucleic acid.

Prokaryote. Unicellular organism which has no nuclei or membrane-bound organelles.

Promoter. Region of DNA involved in binding of RNA polymerase and the initiation of transcription.

Proofreading. The ability of a DNA polymerase to correct errors in nucleic acid sequences during complementary strand synthesis by the excision of incorrectly incorporated nucleotides. Proofreading is accomplished by a 3′ to 5′ exonuclease activity associated with the polymerase.

Protein. A naturally occurring macromolecule composed of linear polymers called polypeptides.

Purine. A two-ring heterocyclic nitrogenous base. **Adenine** and **guanine** are the purines found in both DNA and RNA.

Pyrimidine. One-ring heterocyclic nitrogenous base. **Thymine** and **cytosine** are the pyrimidines found in DNA, and uracil and cytosine are the pyrimidines found in RNA.

Quantitative PCR (QPCR). A variant of standard PCR leading to the enumeration of target sequences.

Ramp rate. Rate of temperature change between set temperatures, as determined by the thermal cycler.

Random amplified polymorphic DNA (RAPD). A PCR-based DNA profiling method to detect DNA polymorphic differences.

Random priming. Method of DNA probe labelling using random sequence oligonucleotide primers and the Klenow fragment of DNA polymerase I.

Reannealing/renaturation. Mechanism by which complementary single strands of nucleic acid associate to form a double-stranded molecule. Opposite of **denaturation**.

Reference standard/reference material. A substance, one or more properties of which are sufficiently established to be used for both apparatus calibrations and the intra- and inter-laboratory validation of analytical techniques.

Repetitive motif. Repeated stretch of nucleotides in a nucleic acid sequence.

Reporter. A species that displays some physical, chemical or spectroscopic property which enables it to be detected. Commonly used for labelling DNA probes.

Restriction endonuclease/restriction enzyme. A bacterial endonuclease that cleaves duplex DNA at specific sequences (**restriction sites**).

Restriction site. The DNA sequence recognised by a specific **restriction endonuclease**.

Reverse transcriptase. An enzyme that synthesises DNA from an RNA template.

Ribonucleotide. The monomer of RNA that consists of ribose-phosphate esters with each ribose unit carrying a base (A, U, G, C).

Ribose. Five-carbon sugar component of RNA.

Ribosomal RNA (rRNA). RNA that is incorporated into ribosomes.

Ribosomes. Small cellular particles composed of ribosomal RNA and protein. Involved in protein synthesis.

RNA (ribonucleic acid). Polymer of ribonucleotides, located mainly in the cytoplasm. Different classes of RNA have distinct roles in the transcription and the translation of the genetic messages into polypeptides.

RNase. A nuclease which cleaves RNA molecules after U and C bases. See also **nuclease**.

r*Tth* DNA polymerase. The recombinant form of *Tth* DNA polymerase.

Secondary structure. The folded structure formed from the interaction of polymeric chains of a nucleic acid or protein.

Sequence. Primary order of nucleotides in a length of DNA or genome.

Sequencing. The method of determination of the primary sequence of a DNA molecule.

Short tandem repeat (STR). Regions within eukaryotic genomic DNA consisting of simple tandemly repeated sequences, usually 2–6 bp in length (microsatellites). STRs exhibit a high degree of length polymorphism due to variation in the number of repeats, and provide a source of highly informative loci for use in the identification of individuals.

Site-directed mutagenesis. The method of introducing predefined base change(s) at a specific site in a target DNA sequence.

Southern blotting. The method of transferring DNA from an agarose gel to a solid support such as a nylon membrane.

Spiked controls. A type of positive control, commonly used in DNA extractions, to check for successful DNA extraction from a particular matrix. Involves the addition of pure DNA or cells to that matrix.

Stoffel DNA polymerase. Form of *Taq* DNA polymerase which lacks 5' to 3' exonuclease activity, due to amino-terminal deletion.

Stringency. Reaction conditions, in particular temperature, salt concentration (ionic strength) and pH that affect the specificity of hybridisation or annealing of two single-stranded nucleic acid molecules.

***Taq* DNA polymerase.** A thermostable DNA polymerase isolated from *Thermus aquaticus*, which is exploited for PCR-based techniques. Possesses 5' to 3' exonuclease activity but not 3' to 5' exonuclease activity (**proofreading** activity)

Template. Nucleic acid from which a complementary strand is synthesised by a polymerase.

Thermal cycler. A device with a temperature cycling block used for performing PCR and related nucleic amplification techniques.

Thermal cycling. The temperature cycling process of template denaturation, primer annealing and extension during PCR.

Thymine (T). A pyrimidine base found in DNA or RNA.

Touchdown PCR. A modification of standard PCR which involves using an elevated primer annealing temperature during the initial cycles and then gradually reducing the set temperature over the following cycles to the predetermined optimal annealing temperature. Touchdown PCR can improve the efficiency and specificity of amplification.

Transcription. The production of mRNA from a DNA template by an RNA polymerase.

***Tth* DNA polymerase.** A thermostable DNA polymerase isolated from the thermophilic bacteria *Thermus thermophilus*, which displays reverse transcriptase activity in the presence of Mn^{2+}.

Uracil (U). A pyrimidine base found in RNA only.

Uracil DNA glycosylase. An enzyme that catalyses the removal of U from nucleic acids containing U in place of T.

UV irradiation. Electromagnetic radiation with a wavelength shorter than that of visible light (200–390 nm), used for sterilisation purposes.

Vector. A DNA molecule, e.g. plasmid, used to clone DNA sequences of interest.

Wild type. The genotype or phenotype of a given organism as found in nature.

Subject Index

www.ingramcontent.com/pod-product-compliance
Lightning Source LLC
Chambersburg PA
CBHW070714220326
41598CB00024BA/3152